The Environmental Career Guide

The Environmental Career Guide

Job Opportunities with the Earth in Mind

NICHOLAS BASTA

John Wiley & Sons, Inc.

New York • Chichester • Brisbane • Toronto • Singapore

Text printed on recycled paper.

In recognition of the importance of preserving what has been written, it is a policy of John Wiley & Sons, Inc., to have books of enduring value published in the United States printed on acid-free paper, and we exert our best efforts to that end.

Library of Congress Cataloging-in-Publication Data

Basta, Nicholas
    Environmental career guide : job opportunities with the earth in mind / by Nicholas Basta.
        p.   cm.
    Includes bibliographical references.
    ISBN 0-471-53416-1 (cloth : alk. paper)—ISBN
0-471-53413-7 (pbk. : alk. paper)
        1. Environmental protection—Vocational guidance.   2. Pollution control industry—Vocational guidance.   I. Title.
TD170.B38   1991
363.7'0023—dc20   91-15587

Printed in the United States of America

10 9 8 7 6 5 4 3 2 1

For the inheritors of the Earth—
the children—especially my son and daughter,
Adrian and Nicole

Preface

Over the past 25 years—and especially over the past 10—environmentalism has grown up. As an ethic of living with the land and its creatures, as a science of understanding humanity's impact on the Earth, and as a profession to which one can devote a career, environmentalism has now become a national institution. This is not to say that all controversies surrounding environmental issues have disappeared—far from it. If anything, the pressures are rising.

What has changed, however, is the belief that environmentalism will "go away" once the current batch of disputes is resolved. Environmentalists used to be a radical fringe in American society. Now, as one public opinion poll after another has shown, environmentalism is the mainstream. Those who believe that the Earth can be exploited straightaway have been relegated to the fringe.

Along with this maturing comes the rising professionalism of environmental work. The number of college-level programs has grown dramatically. The degree of specialization in professional work has increased. Some jobs in industry, consulting, and the leading research institutions can be tremendously lucrative. "Green-collar workers" is the new term for these professionals who are devoted to the betterment of the Earth. By some estimates, there are about three million such workers in the United States.

This book explains what these jobs are, who the employers are, and how best to prepare for green-collar work. If you are committed to improving the environment, and if you have experience or education to bring to the table, the green-collar work force is waiting for you. If you also are ambitious, there may be no other place to go where you can advance your career while doing work that is important to you. This book will help you make those decisions and then find the right place in the green-collar work force.

Chapters 1 and 2 provide an introduction to and capsule history of environmentalism in the United States. A rich tradition of activism, scientific research, and intense conflict characterizes U.S. environmental history. Knowing the background of today's environmental controversies will help you make choices about your future direction.

Chapters 3 and 4 provide an overview of today's dynamic environmental businesses. This is cast in light of the green-collar work force and includes such elements of the businesses as insurance and real estate, occupational safety and health, communications and advertising, and teaching. Key laws, some of which have created billion-dollar industries almost overnight, are reviewed as well.

Chapter 5 provides details on science and engineering, public administration, FIRE (finance, insurance, and real estate), law, communications, and teaching—the pivotal professional concentrations in the green-collar work force. Included are capsule summaries of 30 of these professions.

Chapter 6 offers some basic career-guidance information on how to go about getting a job. One appealing aspect of environmental work is that the availability of entry-level training, through internships, is relatively high. This and other points will be explored.

Chapter 7 is a case study, of sorts: the story of how chlorofluorocarbons (CFCs), a miracle product in the 1930s, came to be banned in the 1980s. Besides being a great story—with a happy ending!—the case provides a clear look at how all the professions and factions of society are coming to bear on environmental issues today. The CFC story has been told in a number of books, but no one, to my knowledge, has looked at it in terms of the professions and skills involved in resolving this trail-blazing international issue.

A special effort has been made to compile information sources; these make up Appendixes A through E. These include state and federal agencies involved with environmental regulation; magazines and other periodic literature, and public-interest and professional societies. Also included is a listing of educational organizations and the degrees and/or programs they offer.

Contents

CHAPTER ONE
Introduction to the Green-Collar World   **1**

The Industry: $2 Trillion and Growing   3
Who Are the Green-Collar Workers?   4
Are You Ready to Join?   6

CHAPTER TWO
Recent History of Environmentalism   **8**

Background: Babylon to Rachel Carson   8
The Industrial Revolution in America   9
Post-World War II   12
Earth Day and EPA, 1970   14
Industrial Accidents Set the Stage   16
Thinking Green Collar   19
Eco-Manufacturing   21
Environmental Police: Government   21
Environmental Activism and Nonprofit Organizations   22
Touching the Future: Teaching   25
A Personal Note   26

CHAPTER THREE
Today's Environmental Business   **29**

Follow the Money   29

The Lay of the Land   30
The Environmental Job Machine   33
The Rule Book of Environmental Laws   37
    The Wilderness Act of 1964   37
    The National Environmental Policy Act of 1969
      (NEPA)   39
    The Clean Air Act of 1970 (CAA)   40
    The Clean Water Act of 1972   42
    The Safe Drinking Water Act (SDWA)   43
    The Resource Conservation and Recovery Act of 1976
      (RCRA)   44
    The Toxic Substances Control Act of 1976
      (TSCA)   47
    The Comprehensive Environmental Response,
      Compensation, and Liability Act (CERCLA, or
      "Superfund")   47
    State and Other Laws   51

CHAPTER FOUR
Environmental Business II   **53**

Private Industry   54
    Pollution Generators   54
    "Green" Marketers   61
    Recyclers   64
    Disposers   67
Government   72
    Federal Agencies   72
    State Agencies   80
    Local Government   81
Business Services   82
    Science, Engineering, and Architecture Consulting   82
    Legal Services, Insurance, and Risk Management   88
    Market Research   90
    Communications, Publishing, and Advertising   92
Travel, Tourism, and Recreation   94
Teaching   96
Nonprofit Organizations   97

CHAPTER FIVE
Environmental Professions   **100**

Spotlight on the Environmental Manager   102
Green-Collar Professions   103
    Agronomist   103
    Air Quality Engineer   104
    Biologist   104
    Biostatistician   105
    Chemical Engineer   106
    Chemist   106
    Civil/Environmental Engineer   107
    Community Relations Manager   107
    Computer Specialist/Database Manager   108
    Contract Administrator   109
    Earth Scientist   110
    Ecologist   110
    Environmental Protection Specialist   111
    Forester   111
    Fund Raiser   112
    Geological Engineer/Hydrogeologist/Hydrologist   113
    Grass-Roots Coordinator   113
    Hazardous Materials Specialist   114
    Industrial Hygienist   115
    Interpreter   116
    Lawyer   116
    Lobbyist   117
    Noise-Control Specialist   118
    Planner   118
    Risk Manager   119
    Safety Engineer   120
    Soil Scientist   120
    Technical Writer   121
    Toxicologist   122
    Water Quality Technologist   123

CHAPTER SIX
Entering the Job Market   **124**

Technical versus Nontechnical Jobs   125

Volunteer Work   128
Government Service   129
Private-Sector Jobs   131
Education   133
The Moral Dimension   136

CHAPTER SEVEN
A *Case Study*: CFCs and the Ozone Crisis   **138**

Introduction   138
Refrigeration Worries   139
New Markets   140
The Question   142
First Reviews   144
On to the Antarctic   146
The Montreal Protocol   148
Industry Adjusts   149
Who the Players Were   151

APPENDIX A
Nonprofit Organizations   **155**

APPENDIX B
Environmental Publications   **163**

APPENDIX C
Federal Addresses   **167**

APPENDIX D
State Agencies   **172**

APPENDIX E
Educational Organizations   **178**

References   **189**

Index   **191**

CHAPTER ONE

Introduction to the Green-Collar World

Who isn't an environmentalist? Whose career doesn't involve the environment in some way?

Most people, if asked, would probably agree that clean air, clean water, wholesome food, and plentiful wildlife are good things. But most people would hesitate to call themselves environmentalists. Similarly, most business managers, while amiably concerned about environmental conditions around the world, believe that the most serious environmental problems belong to some other company or some other industry. And most business managers also believe that the solutions are to be found among the "experts"—scientists, engineers, and policy makers—whose job it is to take care of the environment.

But consider these examples, drawn from the headlines in recent years:

- California's Silicon Valley electronics industry prided itself on clean, quiet factories radically different from the smoky, dusty steel mills and machine shops of the midwestern Rust Belt. What a surprise, then, when it became clear during the mid-1980s that Silicon Valley was polluting its groundwater with carcinogens, pumping out gases hazardous to the atmosphere, and injuring its workers by exposing them to harmful chemicals.

- Mortgage bankers, accustomed to writing up loans to develop shopping centers and commercial skyscrapers, suddenly came to realize that the buildings they were dealing in contained asbestos (a life-threatening mineral), that the ground some of these buildings stood on was polluted from decades-ago industrial activity, and—worst of all—that they themselves would be responsible for cleaning up those sites.

1

■ In the Great Lakes regions, home to some of the finest fishing in the world, sport fishers looked around at the open pit mines, the manufacturing plants producing toxic wastes, and the land developers carving up the wilderness and decided to do something about it. These fishers are now considered as among the most successful environmentalists in the country.

These examples could go on and on. With each passing year, more and more people are coming to realize that, in their work or in their personal habits, they have an effect on the environment. Some are concerned with what humanity is doing to the planet but don't act on that concern because they aren't "environmentalists." Others, however, can see that working for the good of the planet can be a successful and even lucrative career.

I call these people *green-collar workers*. In today's service-driven economy, the distinctions between the traditional "blue-collar" worker and the traditional "white-collar" manager have faded. The computer keyboard a factory worker is tapping at to run an automatic process might be the same one an office manager is using to compile next year's budget. The green-collar workers, however, are only just coming into focus. These are the professionals who are adapting their chosen line of work to the needs of the environment. Green-collar workers are finding that they can combine their personal interests in the environment with the demands of business, government, or other social organizations. In fact, by being aware of environmental concerns, these workers are at the forefront of change in their professions and are finding that their careers are advancing faster for just that reason.

As little as a decade ago, a gulf existed between the "polluters" and the "environmentalists" that was easily as large as the gulf that ever existed between blue-collar and white-collar workers. Things are changing today. Business managers in industry are doing more for the environment, in terms of reducing pollution or making environmentally sound products, than any number of save-the-Earth campaigns. Scientists accustomed to doing abstract, other-worldly research in their specialty are now finding that their knowledge is directly applicable to pollution issues. Conversely, environmentally radical public-interest groups are doing business plans and economic analyses worthy of a Wall Street banker, showing how industry can save costs and protect the environment by conserving energy and materials.

The historical clash between "pro-growth" and "pro-environment" perspectives is not only fading, it is being turned upside down. This confrontation is being inverted by those who are showing how businesses and the economy can grow because the two are doing a better job environmentally. Consumer-goods marketers are finding how new products—verified for their environmental friendliness—are winning sales from savvy consumers. Government regulators are finding how market incentives such as salable pollution permits (which can be traded among buyers and sellers almost as a stock certificate is) could speed up environmental improvement. In contexts like these, who isn't an environmentalist?

Today's green-collar job market needs people who connect their personal values with the environment and the future of the Earth. And it needs people who have special expertise in science, business management, public policy making, and multimedia communications. "It's not enough anymore to say that you are committed to the environment," says Edward Demos, a college professor at the University of Denver and a city environmental official in that city. "What do you bring to the table? What problem-solving skills do you have? This is what the job market is looking for."

And for job seekers with the right mix of experience, knowledge, and commitment, the financial rewards can be eye-popping. Government staffers, traditionally in a low-paying salary structure, can double or triple their income on the outside. Academic scientists can turn entrepreneurial ideas into million-dollar investments. Consulting engineers who started in a garage can be building multinational conglomerates. And somewhere in America today, I feel certain, there is a small-business manager, who started a company to do something good about a local pollution problem, who will be running the equivalent of an IBM or Exxon of the twenty-first century.

The Industry: $2 Trillion and Growing

Every projection of the cost of solving an environmental problem in the past 20 years has been underestimated. The final bill always comes in higher. My estimate—which also will problably turn out to be an underestimate—is that between now and the end of the century $1 trillion will be spent directly on the environment in the United States. Revenues associated with, but not directly spent for, the environment, such as the development, manufacture, and distribution of environmentally sound consumer products, combined with indirect expenditures, could easily double this figure.

Two trillion dollars? That's unbelievable.

But it's not farfetched and is not even a very bold estimate. The U.S. Environmental Protection Agency (EPA) calculated in 1990 that the country spent 2 percent of its gross national product (GNP) on environmental protection and remediation—roughly $100 billion per year. Ten years of that level of expenditure (not even assuming growth in GNP) adds up to $1 trillion. These expenditures, however, do not include such activities as the recycling industries, which obviously are a benefit to the environment. Pollution prevention—the development of new products and processes that do not create pollution—is not included, because generally, once a problem has been solved, it disappears from the list of "environmental" activities and is counted in the lists of normal economic activities. (Preventive medicine has the same problem: How can a doctor justify charging you for keeping you from getting sick?)

The cost of educating workers for environmental work—college, primarily—is kept on a separate ledger. The preservation or restoration of wilderness is kept on another ledger. The expense of writing and selling this book (which I, for one, would count as environmental work) is kept on another ledger. Finally, although the discussion here is limited to U.S. activities, it is undeniable that America's expertise in environmental protection is being exported abroad in the form of projects or technology transfers that are paid for by other countries. What portion of that business should be counted as U.S. economic activity?

Thus, $2 trillion is not an excessive figure. Whether or not you accept that figure, it is certainly true that the rate of growth it is experiencing currently is much higher than the U.S. economy as a whole. The signs of this are all around: an EPA budget that grows twice as fast as the federal government's, the explosive growth in environmental law practices, the jump in the number of schools offering environmental programs, and so on.

Another way to demonstrate what is driving the environmental business of today is to look at Figure 4 (in Chapter 3), which is simply a chart of federal legislation plotted against time. The upward slope of this line is a textbook example of geometric growth; it is literally an explosion of activity. I do not suggest that there is a one-to-one relationship between the number of laws passed and the money being allocated to environmental work, but there usually is some kind of financial connection. Some pieces of legislation, such as the latest version of the Clean Air Act (passed in November 1990) are expected to require the expenditure of $20 billion annually.

Today, all this points to a large green-collar work force that is bound to grow larger in the near future.

Who Are the Green-Collar Workers?

It's time to stop thinking about environmentalists and industrialists on two extremes with the rest of the population vacillating in between. Think instead about who is having a positive effect on the environment or on social practices that affect the environment:

- Government officials who regulate polluting industries, allocate society's resources to cleaning up past environmental mistakes, and help organize government-industry cooperatives to solve pressing environmental problems.

- Business managers, marketers, and economists who are finding new ways to capitalize on better environmental quality and who are helping to transform the polluting methods of manufacturing and distribution.

- Engineers and architects who design nonpolluting manufacturing processes or products that make better use of the Earth's resources than many of today's throwaway consumables.

- Scientists, ranging from atmospheric chemists to tropical jungle entomologists, who are finding out more about the Earth, helping to show where human activities are causing harm, and devising better ways to conserve the Earth's resources.

- Lawyers who defend the Earth in the courtrooms across the country, help draft new legislation that clears away obstacles to improving the environment, and help make polluters pay dearly for their environmental insults.

- Risk managers and public safety and health specialists who are showing how the methods of doing business and allocating risks that were developed to allow society to respond to immediate disasters (fires, floods, and the like) can now be applied to slow-to-appear environmental disasters.

- Communicators, ranging from policy advocates to public-interest groups, to editors of technical journals, to journalists that help explain arcane technology in everyday terms.

- Teachers, who are helping to create an environmental ethic in a new generation of youth and who provide the critical technical training that today's environmental businesses are seeking.

In all these cases, it makes little sense to attempt to identify pro-environment or pro-growth individuals. There will be professionals representing every shade of the spectrum among the various organizations that come to bear on an environmental problem. The time for ideology—both of the save-the-Earth and the Reagan-style envirobashing varieties—has passed.

What these professions have in common is the development of a body of specialized knowledge that can be brought directly to bear on environmental problems. In some professions (notably engineering), this body of knowledge has been centuries in the making. In others (such as risk management), the basic textbooks are still being written.

The study of the environment is the study of science. One way or another, these professions involve scientifically derived knowledge. Either an individual has to understand science in order to make a meaningful contribution or has to react (in the political, business, or social arenas) to what the science says. However, this does not mean that only people with a science education should be interested in the environmental field; nothing could be farther from the truth. "Human beings are the instruments of scientific progress," writes Garrett Hardin, one of the more famous scientists in the environmental arena, in *Filters Against Folly*. "Technical results are often difficult to understand, but the peculiarities of the instruments—fallible human beings—are both fascinating and understandable by all." More to the point, he notes that science is but one filter through which people view reality. The scientific filter needs constant correction by other, nonscientific filters.

On a more practical level, many green-collar workers are finding that they can cope with the level of scientific debate going on around them as long as they are open to learning and have the patience to study the issues. "For the work I do, I think it's beneficial that I don't have a science degree," says Robin Woods, a public affairs specialist at EPA. "My job is to translate the work done by scientists, engineers and policymakers here into terms that are meaningful to the common person. Most scientists are unable to do that effectively." And before you say that Ms. Woods' work is "only" public affairs, you should realize that the EPA policy makers regularly consult with her for feedback from the press and public to see how environmental decisions are being received. "A rule or policy that no one understands or cares about is a bad one," she says.

And, as Ms. Woods herself demonstrates, the green-collar work force depends on women as well as the men one might expect from such male-dominated professions as science, engineering, or business management. "A lot of women were new entrants to the work force as the environment became a big issue in the 1970s and 1980s," notes Gail Brice, head of an environmental-business consulting firm in California. "Those women now represent a substantial portion of environmental management around the country; you could even say that an old-girls' network has formed."

Are You Ready to Join?

So, if you're committed to improving the environment, if you have an expertise or body of knowledge to bring to the table, and if you have scientific training (or the willingness to deal with scientific issues), the green-collar work force is waiting for you to join. Finally, if you're all of the above plus being ambitious, there may be no other place to go where you can advance your career while doing work important to you.

This book will help you make those decisions and then find the right place in the green-collar work force.

CHAPTER TWO

Recent History of Environmentalism

"The environment" as a field of work or study has only recently come into being as a separate, specialized profession. But this is not to say that the concerns raised by environmental issues today—human health, nature preservation, minimizing the damages of human activity to the Earth—are brand new. Today's environmentalism has had many forebears and that heritage shapes the activities going on today.

This chapter will provide something of a capsule history of environmentalism in the United States. Knowing where it has come from gives some perspective on where it's going. The newness of professional careers in the environment creates a large number of precedents, traditions, or simple biases for and against certain types of academic or professional preparation for careers, and knowing about these trends will help you in selecting college programs or provide some insight into where to derive the work experience that will get your career moving.

Background: Babylon to Rachel Carson

For most of recorded history, environmental problems were usually a conflict between human actions and humans themselves. Wherever people have gathered, the most obvious environmental problem has been sanitation. When urbanization evolved in the period starting around 2000 B.C., public sanitation arose as an expression of civic rule. Even as far back as ancient Babylon, sewage systems were a part of the urban structure. Sanitary laws keeping wastes away from settlements were written into the Mosaic code around 1500 B.C.; Jerusalem had its own water-supply and sewage system.

The ancient Greeks, and to a much greater extent the ancient Romans, established many sanitary codes in their city-states. Roman aqueducts, of course, remain a part of the vista of many cities ringing the Mediterranean. The first municipal dump, as it were, is recorded in 500 B.C. in Athens. Plato, it is said, wrote an essay decrying deforestation in the vicinity of Athens.

Ancient societies were agrarian in nature, and agricultural policies arose primarily to ensure that one year's satisfactory crop would be followed by another the following year. Crop rotation, to maintain the fertility of soil, was established by 50 A.D. by Roman farmers and apparently in even earlier times by Chinese and Indian farmers. Extensive land tillage was the norm in Chinese cultures around the fourth century A.D. and in Incan cultures around 1000 A.D. As some areas became heavily irrigated, the problem of salinization of the soil affected parts of India and the Middle East.

It is believed that much of the inner European continent was settled by the same slash-and-burn practices that are so criticized in the Amazon Basin today. The most serious environmental problem confronting the medieval Europeans was in their cities, where poor sanitation aided the spread of the plague. Notwithstanding the large cities, these cultures were still primarily agrarian in nature.

Many historians peg the rise of the Industrial Revolution, in the mid-eighteenth century, as the era when serious environmental problems began to appear. But the evidence is that many problems arose sooner. Due to the needs of iron makers who used charcoal derived from wood, England had exhausted many of its forests during the late sixteenth century, and a law restricting timber use was put into effect during Queen Elizabeth's reign. In England and other parts of Europe, the depopulation of native species for meat or sport led to royal decrees against hunting in certain forests; thus, poaching was born. Old European ports were being filled with a steady flow of garbage from their populations. As the Industrial Revolution took hold, environmental problems like these intensified. Coal began to replace wood for many fuel and energy uses. With manufacturing on the rise, the problems of industrial hygiene and air and water pollution appeared, many of which are with us to this day.

The Industrial Revolution in America

In North America, a new world lay open to the adventurers from Europe. The entire nineteenth century represented a gallop over the continent until it was settled from coast to coast. Many of the forests of the Great Lakes regions were clear-cut to supply the burgeoning demand for lumber. The settling of the Great Plains, retarded by the lack of wood, sped up as barbed wire became a fencing material and the change in plant culture, from open plains to ranchlands and then to farmlands, began. Even as this was occurring, the desire to preserve some parts of the countryside cropped up. Abraham Lincoln took time out from attending to the Civil War to sign into law a bill protecting Yosemite Valley in California in 1864. Simultaneously, many of the great urban parks of the cities of the northeast were being built; apparently, the need for parkland touched a common cord among many Americans.

Yosemite was followed by Yellowstone in the 1870s. The great naturalist John Muir, founder of the Sierra Club (in 1892), intensified his campaigns for preservation of western wildernesses at about the same time that Frederick Jackson Turner, the historian, commemorated the end of the American frontier and the conclusion of America's "manifest destiny." All these forces came to bear on the pivotal role of Theodore Roosevelt, president from 1901 to 1908. Reevaluations of the national parks system were instigated by the plundering of the federal reserves by private interest; laws regulating the purity and efficacy of foods and drugs were passed.

During the course of the nineteenth century, the Industrial Revolution took hold in the United States, as it had earlier in Europe. One manifestation of this was the development of modern water-supply systems, beginning with the city of Philadelphia's, which was built during 1800 and expanded and rebuilt a few decades later. A "Sanitary Movement," as it was called by public health authorities, gained momentum in many of the larger cities of the East during this century. Key components were an enlarged supply of fresh water and better control over waste disposal. An appalling number of plagues and fever breakouts affected many cities during this era. Thousands died regularly every summer, especially in southern cities.

Already, Pittsburgh, the archetypal "smoky city," was experiencing air-pollution problems as a result of the combustion of dirty coal. Lacking a correct understanding of the cause of diseases such as tropical fevers, medical professionals believed that miasmas from sewage and rot were the source of the fevers that so traumatized cities of the time. Paradoxically, Pittsburghers felt that their smoky air kept them healthier by preventing the miasmas from forming.

A contemporary of Theodore Roosevelt (and cutting the same rambunctious reformist figure), George Waring, came to be known as the "apostle of cleanliness" at the end of the nineteenth century, when he was appointed street-cleaning commissioner of New York. "At the core of his views on sanitation was an adherence to Hippocrates' adage 'Pure air, pure

water, and a pure soil' and the belief that cleanliness was a gauge of civilization. 'There is no surer index of the degree of civilization of a community,'
he said, 'than the manner in which it treats its organic wastes,'" cites Martin
Melosi in *Garbage in the Cities*, a history of urban sanitation.

Among his many reforms was the establishment of "source separation"
as a means of reducing the volume of garbage by segregating recyclables
from the mix—this being one of the essential elements of modern-day solid-
waste disposal. He also raised the pay of sanitation workers (to $2 per day),
dressed them in white uniforms, and began a Junior Street-Cleaning League
for the youth of the city.

The picture of the environment between the dawn of the twentieth
century and the end of World War II is one of growing industralization, the
emergence of the automobile, and the concomitant need for highways.
There were few national, federally funded or mandated environmental actions; most issues were resolved (or left to fester) at a local level. One exception to this was the Rivers and Harbors Act of 1899, which, among other
things, prohibited the discharge of wastes into navigable waters. That act also
established the role of the U.S. Army Corps of Engineers in monitoring and
developing water resources, a role that continues, with no little controversy,
to this day.

The Depression brought such projects as rural electrification, as exemplified by the enormous Tennessee Valley Authority, which combined flood
control with electrification and industrialization. At the same time, federal
funds were made available for the grand national dams in the West and expansion of the national parks systems. This era, too, was the dawn of the meteoric
rise of the automobile. Laws relating to the use of autos and of funding and
taxation mechanisms for building new roads were established. Perhaps the
most dramatic changes, though, were those of a social rather than legislative
nature. The horse disappeared as an urban-pollution concern, only to be replaced by the air pollution caused by internal-combustion engines. With the
passing of the horse-based mode of transportation, the horse-breeding and
feeding industries had to change as well. (Rumor has it that the breakfast cere-
als introduced by Kellogg and others at this time were partly motivated by a
desire to find new markets for what had been essentially horse feed.)

The pre-World War II era saw the growth of new social and professional concerns and organizations to address them. Today's Air and Waste
Management Association (Pittsburgh) was founded in 1907 as the International Association for the Prevention of Smoke. It is perhaps no accident that
the extinction of the passenger pigeon (the last recorded bird died in a zoo in
1914) happened at about the same time as the birth of the National Au-

dubon Society. As big land-development issues such as dams, power plants, and surface mining arose during the Depression, the Wilderness Society was formed in 1935 and the National Wildlife Federation in 1936 to protect against the depletion of wildlife and wilderness areas.

Post-World War II

After World War II, U.S. industry underwent an order-of-magnitude change in size, productivity, and types of products. The petrochemical industry, boosted by the wartime search for alternatives to natural materials such as rubber and silk, unleashed a flood of new synthetic materials and substances. Electricity from nuclear energy became a reality as a result of the "Atoms for Peace" program started by the Eisenhower administration. Arguably, the most dramatic successes of the era—the Salk vaccine for preventing polio and the sulfa drugs for controlling infections—led to a belief that human ingenuity could conquer any perceived social or medical ill. Pesticides such as DDT were sprayed wherever insects were a problem. Floods in river basins were cured by channels and locks and, where possible, a dam that could be used to produce hydroelectric power. A national highway system was constructed and the number of automobiles in the country exploded.

Although there had been a growing number of studies and federal surveys of the health of nature in the United States, no one was quite as successful as the marine biologist Rachel Carson, whose book *Silent Spring* was published in 1962. "Her book has often been cited as perhaps the most influential single factor in creating public concern about the future of the world's ecology," wrote *The New York Times* in a twentieth-anniversary assessment of her work. "It was Rachel Carson, many people agree, who initiated the modern environmental movement."

Several themes are stressed in Carson's book that remain vital issues today. First and foremost, she criticized the indiscriminate spraying of pesticides (including DDT, dieldrin, carbamates, and others) in agriculture and to control tree disease; this spraying killed a broad spectrum of insects and other life forms in addition to the targeted species, destroying the ecology of the area. She pointed out that many of these compounds are carcinogenic to humans; in effect, we were poisoning ourselves. Finally, she pointed out that even as some of the target species for which the pesticides were intended (mosquitoes, especially) were developing a tolerance to the pesticides, there existed a host of biological pest-control methods. Indiscriminate pesticide spraying was a no-win proposition all around.

A federal commission agreed with most of Carson's conclusions a couple of years later, and DDT was banned altogether a decade later. This also is the time when marshes and wetlands on coasts and in river basins stopped being perceived as a mistake of nature, to be drained and paved over, and started to be perceived for what they are: a vital component of complex ecosystems.

By then, newly energized environmental groups and nature lovers were stopping highway construction projects in midstream, fighting for the preservation of more nature parks, and seeking greater control over hazards in the workplace and air and water pollution. The Environmental Defense Fund and the Natural Resources Defense Council, two legal powerhouses in the environmental movement, were formed in 1967 and 1969, respectively. In 1969, the National Environmental Policy Act was passed, and the following year, the U.S. Environmental Protection Administration (later, Agency) was formed. In the same year, the Occupational Safety and Health Act was passed to protect workers from hazards. Laws pertaining to environmental safety and nature preservation have been on a steep curve ever since (see Figure 4, Chapter 3).

Although Carson's book was six years in the writing and created a sensation when published, she wasn't working in a vacuum. Most environmental improvement efforts of the time were of a local or regional nature, tailored to the needs of the specific areas. Carson helped make the environment a national issue, worthy of the attention of Congress and the White House.

A good example of local concerns affecting the environment could be found in Pittsburgh during the 1950s. Citizens there had become accustomed to a smoky fog of dust and polluting chemicals for years, to the point where the city's long-term prospects were "blackened." In the early 1950s, an air-pollution crisis in the nearby town of Donora, Pennsylvania, in which dozens were killed suddenly by a combination of smoke (mostly from coal fires) and stagnant weather dramatically spotlighted the crisis. Civic leaders began a Pittsburgh Renaissance campaign that changed the appearance of the city by the 1960s. The cities' turnaround is now being looked on as a model for cleaning up Eastern Europe.

Similar campaigns were initiated to clean up rivers in Ohio, to return Lake Erie to health, and to put the brakes on rampant development in unpopulated areas. Public health officials sought to eradicate endemic diseases, with the very large difference that in these years, many communities' health problems were caused by industrial pollution, rather than mosquitoes or bacteria.

Earth Day and EPA, 1970

Along with the governmental and institutional reforms that led to the creation of the Environmental Protection Agency, changing social mores led to a widespread grass-roots campaign for the environment during the 1960s, which culminated in Earth Day, 1970. In those years, overpopulation of the earth, the dying off of many species of plant and animal life, the existence of cancer-causing synthetic chemicals, and air pollution from burning fossil fuels were the overriding issues. Disastrous oil spills occurred on the beaches of Normandy and California in the late 1960s. Worries over nuclear weapons combined with worries over nuclear power. The "dehumanization" of the workplace concerned social scientists who were troubled by the growing influence of computerization and assembly-line methods of manufacturing.

During the 1970s, the impetus for improving the environment waxed and waned, as other critical social and economic issues buffeted the nation. An oil embargo by the Organization of Petroleum Exporting Countries (OPEC, which is primarily composed of Middle Eastern Islamic countries) dramatized the growing dependence of the United States on oil. Poor economic performance during the decade, highlighted by nearly out-of-control inflation, hampered the establishment of new environmental policies, as industries contested the cost of implementing pollution-control measures.

Even so, a near tidal wave of environmental laws was passed on the national level; many of these were mirrored in state-based actions. The Clean Air Act and Water Pollution Control Act set the basic strategies for confronting environmental ills: The federal government would provide funds, complemented by state monies, to build public works to improve environmental quality. At the same time, regulations enforced by EPA and OSHA would compel industrial polluters to control emissions, to account for the environmental impact of their products, and to integrate their activities with the overall environmental quality of the regions in which their plants were located.

One of the ironies of this first wave of legislation and regulation was that it caused pollution to be diverted in new directions. When particulate matter from the combustion of coal, for example, is captured by filters, a vast amount of dust and sludge is created—a solid waste that must be disposed of. Environmental regulation begat more regulation, as the pollutants being controlled by one set of rules were found to be uncontrolled when converted into another type of pollution.

Also during the early 1970s, there was long-term debate over the capability of the world to sustain the economic growth that was occurring all over it. This debate was first engaged in a computer study (and later, book) called *The Limits to Growth*, commissioned by an informal group of leading industrialists and political leaders named the Club of Rome. The thesis of the report was that western-style industrialism was beginning to run into ultimate limits to the minerals, fossil fuels, and other nonrenewable resources on which it depends. This report created a sensation when first published— and the clamor became even louder following the OPEC oil embargo—but has subsequently been roundly discredited. Alternative theorists, positing that resources can simply be used and reused, or that industrial societies can compensate for the absence of one material by substituting another, have come forward. However, the issue the Club of Rome raised—how the world's resources are used—has not disappeared.

The OPEC oil embargo, the resulting price rise in energy, and the rising tide of environmental concern have had a major impact on energy policy and energy usage in the United States. Energy and power production—oil wells, coal mines, and utility plants—are among the most significant sources of environmental pollution. Once a natural resource like oil or coal is turned into a useful form of energy, such as electricity, gasoline, or cooking gas, another group of pollution problems sets in. Most significant among these is automobile exhaust, which blankets many cities in a smoggy cloud of health-affecting gases. The smokestack emissions from coal-burning power plants are the primary source of sulfur compounds in the air, and subsequently, the generator of acidic rain and surface waters that threaten lakes, streams, and forests.

By the early 1970s, a new form of energy production cast a broad shadow on the U.S. environment: nuclear power plants. These plants, first designed in the late 1950s and constructed during the 1960s, have the advantage of producing no air emissions during normal operation. But the catch is that the risk of abnormal operation, that is, a meltdown of the radioactive core of the plant, is monumental. A release of such radioactive material could blight an entire region for generations. In addition, the normal operation of a nuclear power plant generates a highly radioactive waste (the depleted fuel rods of the plant) that can last thousands of years. Where to put these wastes became a nagging issue as the number of power plants increased, and remains so today.

For these reasons, energy policies have become inextricably intertwined with environmental issues. The U.S. Department of Energy (DOE) maintains extensive research and policy-making resources for deciding environmental issues. In the late 1970s, federal legislation raised the fuel efficiency of automobiles by nearly a third, resulting in a drop in fuel demand.

DOE and EPA also mandated the use of pollution-control devices on automobiles, power plants, and oil refineries. Today, more than a third of the multibillion-dollar cost of building and running a power plant is tied up in pollution controls.

Industrial Accidents Set the Stage

In 1978, an out-of-control situation at a nuclear power plant near Harrisburg, Pennsylvania—the Three Mile Island plant—traumatized the utility industry. While the actual plant upset was contained within the building— the emergency backup systems worked—there was a brief period of time when the plant operators literally did not know what was going on. That lack of control was the damning factor in turning public opinion against nuclear power. No new nuclear plant has been ordered in the United States since; the subsequent plant upset and radioactive release in Chernobyl, USSR in 1986 only further confirmed this distrust.

A key law in the issues of resource use and pollution control was the 1976 Resource Conservation and Recovery Act (RCRA, pronounced "rickrah" by those who deal with it). In a sense, RCRA represents a culmination of the foregoing environmental laws. As originally conceived, RCRA would serve the double purpose of reducing waste pollution by helping create a situation where wastes could be reused, when possible technically; at the same time, the disposal of these wastes would be more closely regulated.

A couple years after the passage of RCRA, another environmental crisis arose, centered around a waste dump in the Niagara Falls, New York, area known as Love Canal. Love Canal, whose existence was by no means unknown, had been used as a dump for organic chemicals since the 1930s by Hooker Chemical Co., headquartered near there, and other firms. The site had been covered with earthen fill, then deeded over to the city, which eventually built a schoolhouse over it and allowed home construction around its perimeter. Eventually, the chemicals began percolating through the soil, both beneath the site and above it. A hard-fought campaign by homeowners in the area, alarmed at the health risks they were exposed to—and by epidemiological studies showing an increase in disease and disability—led to the declaration of emergency relief by the state and the federal government. The neighborhood was depopulated and shut down, pending a cleanup and further study.

Love Canal was only the first of a flood of similar ecological disasters around the country. The town of Times Beach, Missouri, was closed down after high levels of polychlorinated biphenyls (PCBs) were found in oil that had been used to spray dirt roads around the town. A "Valley of the Drums" in Kentucky had upwards of 100,000 drums of waste; another, in Hardeman County, Tennessee, had tons of waste from pesticide manufacture. A rather inflammatory term, "toxic time bomb," is not erroneously applied to dumps like these, since wastes that are dumped or buried only show up years later in the water drawn from wells in the vicinity of the site. These waste dumps led to the passage of yet another law, the Comprehensive Emergency Response, Compensation and Liability Act (CERCLA). This law is often called "Superfund" because it established a tax on oil and chemical products, the revenues of which would be used to clean up abandoned dumps around the country.

In 1984, another industrial disaster occurred that has had repercussions to this day. In Bhopal, India, a U.S. company, Union Carbide Corp., together with the Indian government, had build a herbicide plant whose products were very useful in rice growing. The product has many usage and safety issues associated with it, but its ingredients, which were stored in tonnage quantities at the plant, were out and out poisons. An accidental release (the company maintains that sabotage occurred) during the night of December 13 caused a cloud of one of these ingredients—methyl isocyanate gas—to be released. The gas is heavier than air, so rather than dissipating upward into the atmosphere, it crept gradually down the streets of Bhopal and through the open windows of the shacks and animal stalls there. Even in very small concentrations, the gas sears flesh wherever water is present—in the lungs, in the throat and nose, and in the eyes.

By dawn, it was clear that the worst industrial accident in history had occurred. The death toll (never absolutely confirmed) was around 2,500; tens of thousands more were blinded or crippled with respiratory injuries. Medical facilities were swamped.

In the aftermath of the incident, the American president of the company was temporarily arrested when he attempted to visit the site. A $470 million settlement was negotiated between the company and the Indian government. (As of this writing, that total could still rise into the billions of dollars.) In the United States, however, the accident made people realize that a crucial gap existed between the producers of toxic chemicals, who have instituted many safeguards within their plants to protect workers, and the communities surrounding the production plants, which may be completely unaware of the presence of toxic chemicals and unprepared for responding to their accidental or intentional release.

This accident, along with the scheduled reauthorization of CERCLA, led to the passage of a group of amendments in 1986. These "community right-to-know" laws specify that emergency-response plans be worked out in preparation for an accident, and that communities be made aware of the volume and type of chemicals being released, intentionally or unintentionally, by factories. Now, in a sense, nearby communities determine what processes factories can operate and what chemicals they can use. At the same time, factory managers are made aware that their responsibility in product management doesn't end when a product leaves the plant shipping docks. Emergency planning must be coordinated with local communities, and the choices for production methods must be made with the foreknowledge that the community will have a say in the matter.

The environmental issues addressed by Superfund and right-to-know rules are of a different order of magnitude than clean air or water issues. Arguably, air pollution stops when cars stop running or factory and utility smokestacks are shut down. Water pollution, by and large, would clear itself naturally when pollutants stop entering streams and lakes. But toxic wastes in the soil can exist for decades. The spread of toxic compounds that can dissolve in water will continue for years to come, poisoning the wells on which nearly half of the population depends for drinking water.

In the past decade, other "immortal" environmental problems have come to the fore. Research initiated during the late 1960s finally proved conclusively that chlorofluorocarbons (CFCs), a class of compounds exceedingly useful for their properties as refrigerants and solvents, were causing the degradation of the ozone layer in the upper atmosphere. The compounds act to scrub ozone out of the atmosphere, thus increasing the amount of harmful radiation from the sun reaching the Earth's surface. An international accord, the Montreal Protocol of 1987, was signed to reduce and eventually cease production of these compounds; their presence in the atmosphere will be noticed for years to come.

Similarly, the presence of large quantities of sulfur oxides in the atmosphere has been associated with "acid rain" and an increase in the acidity of certain bodies of water in the United States and Europe. Once a lake or mountain forest acidifies, many life forms die, upsetting the balance among species. New air-pollution rules signed into law by George Bush in late 1990 will dramatically reduce the emissions of these sulfur oxides.

An even more far-reaching problem is the so-called greenhouse effect, whereby radiation from the sun is held more completely under the Earth's atmosphere because of rising concentrations of a variety of gases in the atmosphere. In this case, about which there is little consensus, the Earth's surface temperature could rise, changing weather patterns around the globe, causing the polar ice caps to melt, and changing the mix of fertile and arid regions. Since all combustion of fossil fuels—oil, natural gas, and coal—generates one of these "greenhouse gases" (carbon dioxide), radical changes in energy use may be in store for the industrialized societies of the future.

A number of environmental theorists have postulated that a turning point is being reached in the tide of human history and its dominion over the Earth. World population is shooting toward 6 billion and will reach 8.5 billion by the year 2025. Species extinctions are believed to be occurring at the fastest rate in the planet's history—even faster than the cataclysm that brought the Age of the Dinosaurs to a close 65 million years ago. While previous pollution crises put the residents in one city or along one river at risk, environmental problems such as chlorofluorocarbons put the entire world at risk.

Even space is becoming polluted with junk. A study from the Congressional Office of Technology Assessment, *Orbiting Debris: A Space Environmental Problem*, noted that while there are about 6,500 objects being tracked continually by the United States, there may be 30,000 to 70,000 bits of debris, one centimeter or larger in diameter, floating in Earth's orbit. (The instrumentation to see such small debris doesn't exist; on the other hand, there have already been recorded incidents of small debris damaging the Space Shuttle.) "If space users fail to act soon to reduce their contribution to debris in space, orbital debris could severely restrict the use of some orbits within a decade or two," OTA concluded.

Thinking Green Collar

There are plenty of alarming reports predicting the end of the Earth as we know it. In a strict sense, humanity will only know for certain that a new cataclysm has occurred after the fact, when there is nothing to be done about it. No one suggests waiting until all the evidence is in before acting. Indeed, a great number of positive actions already are being taken. Many individuals, feeling certain that the crisis point has been reached, are making changes in their own life-styles and attempting to influence the life-styles of the people or society around them. I salute their determination, while reserving my doubts whether the ultimate emergency is at hand.

This book is for the rest of us: People who can see that problems are at hand and are growing to serious proportions, but who want to do something about them within the context of laws, institutions, and organizations (whether businesses, nonprofit organizations, or governments) that exist today. I call this approach "green-collar work."

Green-collar work is similar to the environmentally related activities of the recent past, but it is performed in a different context. Specifically, it is work performed by exercising a body of knowledge (science, technology, and policy) while recognizing and demonstrating the value of the exercise to the rest of society. It exists in business and commerce, in civil institutions and nonprofit organizations, and in social agencies.

Industrially sponsored actions to clean up the environment in the past were done figuratively at the point of a gun. If a company didn't stop polluting a river or tearing up a landscape, it would be shut down by force of law. Businesses today are finding, much to their amazement, that there are opportunities to make money (which is, after all, the only reason for the business to exist) by being environmentally conscious, in fact, by operating to the benefit of the environment. Many products are being marketed today as "green" ones, but too many of them are green in name only, not in fact. A compact fluorescent light bulb, on the other hand, does by its nature benefit the environment by conserving energy when in use and by conserving materials because it lasts 15 times as long as a typical incandescent bulb. It doesn't need to be labeled "green" in order to succeed in the marketplace. (What it does need, however, is an economy where the true cost of producing electricity, including the cost of damage to the environment, is combined with the true cost of producing an inefficient incandescent light bulb. When that occurs, the higher purchase price of the compact fluorescent would disappear.)

In order for businesses to succeed in this new environment, a high level of expertise is required. Inventors such as engineers and scientists must be able to create products that work better, thereby conserving resources and protecting the environment. Marketing managers must be able to analyze the blizzard of new regulations affecting existing products and to predict the future course of society in order to deliver new products when they will be needed. The financial and insurance industries must be able to demonstrate, in terms of cold, hard cash, how environmentally friendly enterprises will perform better over the long term, relative to environmentally ignorant companies today. Communicators are needed who can translate technical and financial environmental advantages into terms the customers, stakeholders, and regulators of businesses can understand and be convinced by.

One of the hallmarks of the late twentieth-century economy of the United States is its profound shift toward a service economy, and it is here that most green-collar jobs will be found. For every one job in manufacturing, there are four other jobs in the service economy. While many lament this trend as evidence of the hollowing out of the U.S. economy, one of the factors that contributes to it is the higher degree of productivity of U.S. manufacturing, which reduces the amount of human labor needed to produce goods. Simultaneously, services for businesses have grown dramatically. It can be said, in an oversimplified way, that the 100 clerks that would have been required by a manufacturer 50 years ago have been replaced by a computer-programming consultant to that manufacturer today.

Eco-Manufacturing

The first of the main sectors of the green-collar work force is employed in manufacturing and business services. There is little to argue that manufacturing is the source of most pollution, either from the natural resources extracted, from the manufacturing processes used, or from the lingering problems in usage or disposal that their products cause.

Business consulting has been one of the hottest growth areas in the U.S. work force over the past decade. And within the consulting field, environmental consulting has been one of the fastest-growing specialties. This consulting ranges from engineering design and construction services to build a pollution-control system to communications specialists who help a business coordinate its legally mandated informing of local communities under right-to-know laws. Business consultants are freebooters; they can consult for (and, implicitly, against) government, business, the media, academia, and other special-interest groups.

Businesses, naturally, are not the only social organization in which green-collar work is performed. The dollars spent on salaries or public-information campaigns today may originally have been created by a business enterprise, but nonprofit organizations and governments have resources, dollars, and, most emphatically, jobs for green-collar workers.

Environmental Police: Government

Government, at the federal, state, and local levels (plus the emerging international level) forms the second component of the green-collar work force. Government regulators, who had been holding the figurative gun to the heads of businesses, are necessary to define the playing field in which environmentally sound enterprises can earn their way. It is less and less desirable simply to dictate the terms under which a business enterprise must function environmentally. Putting a business out of operation because of environmental problems may saddle the state or locality with a polluted dump, which must now be cleaned up at taxpayer expense. Banning the production of a chemical may serve to eliminate its presence in a river, but unless that chemical is banned in the rest of the country, the environmental injury has simply been transferred to another location. Or, if its production ceases throughout the country, better be sure that it simply hasn't been transferred across the border, to another country's jurisdiction. Again, the environment of the Earth is the loser. Is it a better path to work with the producer of the chemical to devise a way to make it without fouling the river? Is the techno-logical expertise at hand, both within the business enterprise and the govern-ment regulatory body, to examine the technical issues this would entail? If the expertise is lacking, can it be found elsewhere and brought to bear on the local problem?

Another very important role of government agencies today is to co-ordinate a variety of interests to a common goal. Consider the problem of municipal garbage. If a city wants to recycle some portion of it, it must be able to demonstrate that a commercially viable volume of the material to be recycled can be delivered, week in and week out, to the business enterprise that recycles that material into salable products. Does a city's own popula-tion deliver these materials voluntarily? Does the city's sanitation depart-ment take it upon itself to deliver them? Does the city run the recycling enterprise itself? Does the city create an incentive for a business to engage in recycling by making itself a valuable customer of the recycled product?

Environmental Activism and Nonprofit Organizations

The third component of the green-collar work force is at nonprofit organiza-tions. These include the well-known environmental groups such as the Si-erra Club or the Environmental Defense Fund, as well as industry-sponsored groups like the Chemical Manufacturers Association or the Hazardous Wastes Treatment Council. The term "nonprofit" is used to unite these two, who normally consider each other blood enemies. They op-erate the same way, essentially using the same policy or action tools; they dif-fer primarily in their sources of funding. "Environmental" groups call themselves "public-interest" groups and tar the other type as "special-interest" groups. Wanting to avoid such value judgments, the term "non-profit" will be used to identify these and a variety of other organizations.

The old Latin expression, *Quis custodiet ipsos custodes?*—Who guards the guardians?—comes to mind when one considers nonprofit orga-nizations. Backed with private donations (which often come with any variety of strings attached), nonprofits seek to keep government honest and to pull it by the nose into areas where it may be unwilling to go. The tools of the nonprofit are the publicity campaign, the in-depth study, and the lawsuit. Some of the nonprofits—the lobbyists—can back their campaigns or lawsuits with political contributions. Nonprofits also seek to inform: to make study of issues, give speeches, and rally support for or against a political or social stand.

Many nonprofits are quick to assert that they, and they alone, occupy the moral high ground, that they are supporting or opposing a measure out of altruistic interest in the common good of society. Would that anyone could so act. Each nonprofit represents a special interest, and the way in which nonprofits seek to influence public policy is forthrightly undemocratic. This is not intended to negate their value, but to identify them for what they are. They serve a useful interest, and when they are supported primarily by the small contributions of individuals (rather than a small number of large corporate contributions or large contributions from a few philanthropists), they act to represent the interests of individuals in the minority in society—a profoundly democratic effect.

Nonprofits have become a permanent part of the political and social terrain, especially when it comes to environmental issues. There was a time (in the 1970s) when it was said, somewhat apocryphally, that the EPA's agenda was set in advance by the collective wishes of the leading environmental groups. While that may have put some issues unjustifiably ahead of others, it did serve to energize the activities of EPA and helped it become a significant branch of the government. Many of the most committed members of the environmental community decry the creeping professionalism of leading nonprofits. Where before one advanced in stature by demonstrating the strongest commitment to the cause, be it pristine wilderness, saving whales, or counting birds, now many of these organizations have experienced lobbyists, fund raisers, scientists, and lawyers who seem to be more committed to the process than the end result. While some observers hailed the accession of William Reilly, formerly the head of the World Wildlife Fund and the Conservation Foundation, as EPA administrator under President George Bush in 1989, others saw it as the final co-option of the environmental movement into the establishment. There should be a place for both the believer (the committed nature lover) and the operator (the technologist or lobbyist).

A special subset of the nonprofit group is the professional societies, which are chartered primarily as educational organizations and restricted from lobbying. (There are a few professional societies that most emphatically do lobby.) The number of professional societies is huge and growing. Old-line organizations such as the American Society of Civil Engineers, the Water Pollution Control Federation, the Air and Waste Management

Association, and others are making room for the new ones, including the National Association for Environmental Management and the National Association of Environmental Professionals. Such organizations may not be in the hot lights of media attention, but they serve an important function for individual practitioners. They offer a means of developing a professional identity and standards of professional conduct. They do indeed educate and formally or informally serve as the conduit to jobs and professional advancement.

There is a well-identified extremist element within the environmentalist community, best exemplified by the Earth First! movement, which argues from *a priori* principles that part of the environment—specifically, the wilderness and wild species of life—must exist independently of human activity. There is no point in trying to justify the value of wilderness in economic or environmental-health terms when you insist that wilderness must be allowed to exist, in pristine condition, regardless of human needs. And, you don't need a degree in philosophy or whatever to make a case for this; you simply need to believe in it with more force than anyone doubting or differing with your view. I don't disagree with the emotional or (to put a label on it) metaphysical view this represents of humanity on Earth, just as I would consider it absurd, in this secular world, to believe that everyone must belong to a specific religion.

Earth First!ers and other environmental extremists perform a vital function in the environmental movement, a function that those around them rely on and that they themselves realize as one of the rationales for going on. By defining the extreme—by forcing people to come to grips with the compromises they live with or the trends that they are abetting—environmental extremists set the terms of the debate. Dave Foreman, one of the founders of Earth First!, writes in his book *Confessions of an Eco-Warrior* that one of the reasons for founding the organization, at the dawn of the anti-environmentalist Reagan Administration in 1980, was "to demonstrate that the Sierra Club and its allies were raging moderates, believers in the system, and to refute the Reagan/Watt contention that they were 'environmental extremists.'" (James Watt, a Colorado politician, was President Reagan's first Secretary of the Interior.) The extremists do not need to get a college degree or form a professional affiliation to make their point; they simply need to possess the fire of total belief. They will not earn a living by doing this; no one likes the bearer of bad tidings. Very often, they will find that no one agrees with them, at least initially. And, sometimes with premeditation, they will be living on society's extremes: destitute or in jail.

Touching the Future: Teaching

A fourth sector of the green-collar work force is the lowest paying, least prestigious, and most frustrating; at the same time, it is the most revolutionary and most effective for altering the future course of U.S. society. This sector is teaching. For years, teaching has slipped lower and lower in social esteem in the U.S. work force. Although there has been an upturn recently, the pay has been minimal, even for credentialed, highly intelligent, and motivated individuals. Nevertheless, the hearts and minds of the Americans of the twenty-first century are in the hands of these green-collar workers. Already, there are signs that a new environmental ethic is taking hold in the youngest generation of Americans. In countless households across the country, harried, confused parents are being instructed by their children in the value and proper practice of waste recycling. On the day before George Bush signed the Clean Air Act of 1990 into law, he presented Environmental Youth Awards during American Education Week. He said, "I think of one man in particular, who won this award last year for launching a recycling program. He stood on this stage and asked me if the White House did any recycling. You talk about pressure! . . . Well, I told him that I didn't think we had a recycling program, but we'd sure be working on it."

There are relatively few "environmental" teaching positions, especially at grade-school levels. Many science and social studies teachers are taking the initiative, in an informal way, to present their coursework in a way that reflects environmental issues. Meanwhile, at the college level and above, there has been a virtual explosion of academic programs pertaining to the environment. These programs, by and large, are not opened in the hopes that some students will show up. Rather, they are being brought into existence by the students' intense desire to study environmental issues. What civil rights, as a cause, was for the college generation of the 1960s, environmental rights are becoming among the college students of the 1990s.

One of the other educational pathways that will be explored in great detail in this book is postprofessional training, in which a certificate of some sort, rather than a degree, is awarded. Again and again, employers and executive recruiters lament the lack of properly trained, experienced specialists in a host of environmental subjects. These range from emergency-response procedures in the workplace to the proper methods for applying to be a government contractor. Green-collar teachers adept at these subjects are in high demand and can earn more-than-respectable salaries.

Business (including business consulting), government and nonprofit organizations, and teaching are the main sectors of the green-collar work force. One of the great questions anyone contemplating a green-collar career will consider is how permanent this field is going to be. Is the "green movement" the fad of the go-go late 1980s, doomed to blow away in the economically constrained 1990s? Much of the next chapter will demonstrate why this will not occur. Nevertheless, it can be said with certainty that the environmental work opportunities of the 1990s will be vastly different from today's, just as today's are greatly changed from those of a decade or so ago. In the 1970s, the composition of environmental-impact statements—a laborious process of determining the consequences of a construction project—occupied the attention of countless business managers. The effort expended in "mere paperwork" was routinely decried. Today, the paper blizzard has grown tenfold, at least. Environmental-impact statements are still being written, but the controversy about them has died down. The environmental battles have moved to other arenas: liability in legal claims for an abandoned dump site; answering the "how clean is clean?" question for an environmental remediation project, defining what a "recycled" or "recyclable" product is, and so on. Change is the constant.

A Personal Note

I am an environmentalist. I am also a journalist, an engineer, a college lecturer, and a career-guidance counselor. We are all colored by our experiences, no matter how objective we believe we are. Here are mine.

I've been working professionally, mostly as a journalist and writer, for about 15 years. If you add five or so years of the time spent pre-, post-, and during college, until a job materialized, this represents a witnessing over the past 20 years. I studied engineering—chemical engineering, to be specific—in college, writing a senior thesis about air-pollution control. I was employed briefly as an engineer for a chemical company, and then began working as a journalist, writing about energy and environmental issues and the professional development of technological careers.

Even in this relatively short span of involvement with engineering, profound changes have occurred. During my education, I can distinctly remember looking at flow diagrams—simple boxed outlines of the steps of a manufacturing process, with lines showing the interconnection of pieces of equipment. More often than not, the web of lines connecting the boxes would end, off to one side, with a little arrow or the representation of a dump truck. The invariable label on this end of the pipe would read simply, "To Disposal." That's it. As far as the engineer was concerned, once the main product was conveyed safely and efficiently through the plant, the job was finished. Whatever wastes or valueless byproducts were produced were simply carted off—to somewhere else.

Engineering is less and less often taught this way, although it probably still goes on in some courses, at some schools. Thank goodness.

To take one step farther back in time, I grew up in a semirural region in western Pennsylvania. While not a mecca for nature lovers, the region had many game reserves and parks and had a sprinkling of small farmers who had been working the region for five generations. Growing up, I hunted, fished, and farmed. Interspersed among those farms were the strip mines of the Appalachian coal belt, which starts in Tennessee, runs a wide swatch through West Virginia, and gives out just below Pittsburgh. Those coal mines, in turn, fed the steel mills that used to make the Ohio River valley famous (or infamous, depending on your point of view). During college, I had the opportunity to see heavy industry—and heavy air, water, and solid-waste pollution—real close, by working summers and vacations in the steel mills and factories of the region.

Two deeply felt realizations stayed with me out of those experiences: one of the factory and one of the country. The first realization was the incredibly wasteful, inefficient, and plainly sloppy manner in which steel mills of the era were run. At that time, nearly half of the steel being melted down from ore or scrap simply recirculated within the plant, having been culled out of the production line at each and every step of the production process. When culled, this new scrap simply went back to the furnaces at the beginning of the process, to be remelted at enormous expenditure of energy and thrown through the line again.

An interrelated realization was the façade of technology: From the outside, all factories, and steel mills especially, look like other-worldly machines or monsters, gasping out smoke and defecating on the ground and in the water around them. An antitechnologist's delight. From the inside, however, a different picture emerges, of men and women running machines, sometimes well and sometimes poorly, making do with the resources at hand. I remember, not in a condescending spirit, that the overall impression was very much like the unmasking scene in *The Wizard of Oz*, where the "great and powerful Oz" is revealed to be a crotchety old man behind a curtain, pulling levers and turning dials. (That last detail is important: Regardless of how huge, powerful and unnatural a technological artifact may be, if you look hard enough, you will find the men and women holding the control levers that operate it. The machine may be supremely powerful, but it still has a human hand guiding it.)

The second realization to stay with me involves the power of nature.

Even in the neighborhood of the steel mills and strip mines, trees grow, animals live, and nature adapts. I've been to my share of mighty national parks with ranges of mountains and splendid isolation. The difference between a pristine wilderness stretching as far as the eye can see and a forest near a mill town that could politely be called a "mixed-use" resource is much, much smaller than the difference between being in nature and being in the artificial environment of a city or shopping mall. I don't bring this up to justify a mixed-use philosophy regarding all parks and forests, but rather to identify the enormously restful and rejuvenating properties of even a little bit of nature. Nature is very much worth fighting for, worth preserving.

CHAPTER THREE

Today's Environmental Business

We product managers think that it's a big deal when we go before the board of directors for ten or twenty million dollars for a new factory expansion. But when the environmental manager sits at the table, the numbers are fifty or one hundred million. Our projects pale by comparison.

—Senior vice president at a major chemical company

Wall Street is intensely interested in the environmental business right now because that's where the potential for the biggest payoff is.

—Business consultant to environmental entrepreneurs

Salary is almost no object when it comes to hiring environmental managers for companies with major pollution problems. The problem is finding people with the right combination of experience and training.

—Administrator for professional management training

Follow the Money

A well-known line from *All the President's Men*, a movie about the downfall of President Richard Nixon due to financial and judicial improprieties, had Deep Throat, the secret news source for the journalists uncovering the scandal, saying simply, "Follow the money." That's a pretty good perspective to keep when considering many of the new professional careers the environmental business is creating today. Green-collar workers, especially those headed for the private sector, can earn $100,000 or more annually, depending on how large a company they join and the level of their expertise. Technology-driven entrepreneurs who succeed in setting up a new company and making it a going concern can cash out by going public or being acquired. These innovators can earn millions, just as similar innovators in computers and microelectronics or biotechnology have done in recent years.

Where is all this money coming from? How long will the spigot remain open? This chapter attempts to provide some reasonable answers to these questions, bearing in mind that the environment business, like most other forms of private enterprise, is subject to ups and downs in its performance. The track record for the environmental industry as a whole—incorporating industrial companies with pollution problems, engineering and scientific consultants, nonprofit organizations, and natural-resource-management enterprises (whether for preservation or development)—has shown a steady rise for over 20 years, with a much-accelerated growth in the past 5 years. Certain specialties within the field, such as air-pollution control, public water projects, and nuclear materials processing, are often subject to extreme swings in federal regulatory policy. But the savviest green-collar workers have learned how to parlay the expertise developed for one environmental problem into another.

Talking about the big bucks to be had these days in environmental work, and then turning to such sacrosanct subjects as rain-forest protection or stopping the slaughter of whales, is bound to strike many people as the essence of crass materialism. However, there are already a number of innovative programs that combine money with nature, such as the establishment of nature trusts in exchange for foreign-debt forgiveness. And there are some new economic theories ("free-market environmentalism," to take one example) that at the very least integrate environmental and economic issues. Remember, it is only when it is certain that society has committed financial resources to the environment in the form of profit-making companies that we can be assured that green-collar careers are here to stay.

The Lay of the Land

Some pollution or environmental damage is obvious—crude oil leaked onto an ocean beachfront, for example. But some is far from apparent: the fertilizer applied to enrich farmland, later running off the ground during a rainstorm and causing a lake to become eutrophic. Thus, pollution and other environmental problems are usually defined in a legal sense; sometimes (not always) there is also a corresponding scientific basis for labeling a substance as a pollutant. The laws of the land, both at federal and state levels, identify the known environmental problems. The presumed problems, about which there is no political or scientific consensus (such as global warming caused by the greenhouse effect) are still being negotiated as proposed new legislation. In addition, there are many types of environmental issues that become a problem only in retrospect. The existence of the Love Canal chemical dump in Niagara Falls, New York, for example, catalyzed the environmental community in the late 1970s and led to the creation of the Superfund law to clean up abandoned dumps. And finally, let's admit it, there are many environmental problems staring us in the face for which nothing yet has been done. A good example of this, all but ignored until quite recently, is the nuclear-weapons complex run by the U.S. Department of Energy for the military. Over 40 years of corner-cutting, falsified data, lack of oversight, and plain ignorance have led to a series of environmental cesspools that will probably cost some $150 billion to fix, according to the U.S. Office of Technology Assessment.

The most important laws that frame the environmental issues confronting the United States today will be reviewed in a subsequent section. For the moment, it is worth noting that when each law is passed, a new market for environmental goods and services is created. Analyses of the cost of implementing those laws (or, conversely, analyses of the markets those laws are creating) tend to be specific to that law. Thus, you will sometimes hear about an engineering-consulting firm being a "good Superfund contractor," or some such legislative qualifier. This section takes a look at the dollars being spent by industry, government, and the individual consumer. Following the money will lead to finding the jobs.

The most authoritative, but perhaps least insightful, analysis of pollution abatement and control (PAC) expenditures comes from the Bureau of Economic Analysis (BEA), a part of the Department of Commerce. It, in turn, depends on the Bureau of the Census (the same folks who count our heads every ten years) for survey data drawn from businesses and government. In their latest analysis (published in November 1990), the preliminary estimate for 1988 PAC spending was about $86 billion. Over the 1984–88 span that BEA analyzes in this report (see Figure 1), PAC expenditures grew an average 6.2 percent per year. A quick-and-dirty projection forward to 1991, using the same 6.2 percent annual growth rate, would put PAC expenditures in 1991 at $102 billion. EPA, conducting its own projection in early 1991, estimated that 1990 expenditures were $115 billion, and that the end-of-decade annual amount could be as much as $185 billion.

These are "hard" dollars, meaning that they are real funds (and not a bookkeeping equivalent of funds or investments) directed toward regulated environmental activities. BEA defines PAC as "goods and services that U.S.

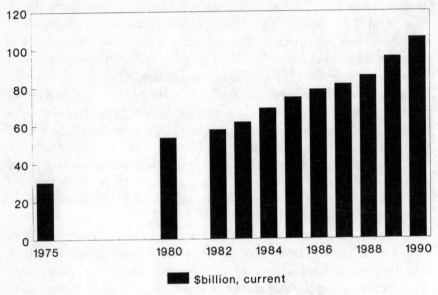

$billion, current

FIGURE I   U.S. PAC Expenditures (*Source*: BEA and industry estimates)

residents use to produce cleaner air and water and to dispose of solid waste."
As is pointed out in other parts of this book, the environmental business is
bigger than the number of automobile tailpipe exhausts and municipal sewage treatment plants that BEA takes such pains to count. Here are some
other numbers, and growth rates, that help put environmental spending into
a job-related context:

- Total recycling industry revenues could top $60 billion by 1994, ac-
  cording to a study from Richard K. Miller & Associates (Atlanta,
  Georgia). This recycling includes paper, aluminum, steel, plastics, oil,
  lead, and tires.
- The medical waste treatment market, currently worth $1.5 billion, will
  grow to $5 billion in 1994, according to Find/SVP, a New York market
  research firm.
- The removal, cleanup, and replacement of underground storage tanks
  (which have leaked toxic chemicals into groundwater) was originally
  estimated at $20 billion over this decade by EPA. Hart Environmental
  Management Corp., a New York firm, estimates that that total will
  double to over $40 billion.

- Alternatives to the landfilling of hazardous wastes, estimated by Frost & Sullivan, Inc., a New York market research firm, cost $10.7 billion in 1989; by 1994, according to a survey, that cost will have doubled to $20.6 billion.

- The State of Florida will have to spend $17 billion over the current and next decades for storm water control, a result of its traditional practice of building ditches in rural areas to handle rainwater flow. As urbanization has occurred, the ditches now convey all the contaminated runoff from paved areas and industrial complexes.

As Everett Dirksen, a U.S. senator of the 1950s and 1960s used to say, "A billion here and a billion there, and pretty soon you're talking about real money." These enormous sums have caught the eyes of many business managers, and not just in the late 1980s. There have been previous jumps in environmental spending; perhaps the biggest jump occurred during the early 1970s when the first batch of environmental laws was passed. This time around, however, it appears that the commitment will last. One reason for this is the legislated mandates to clean up manufacturing operations; even when the federal government becomes lax in enforcing these laws, states are ready to step forward to keep up the pressure. A second reason is the will of the American population; fully three-quarters of the people say that they are willing to pay more for environmental cleanliness, regardless of the expense or human cost when jobs are lost because a polluting plant is shut down.

The Environmental Job Machine

This last point—that jobs are sometimes put at risk by tighter environmental rules—bears exploring, because traditionally, it had been used as an argument against those rules. Labor unions, for example, often lobbied against some environmental regulations because of the pressure those regulations put on their employers. And, as recently as 1990, the job question came up again in the intense debate in the Pacific Northwest over the preservation of "old-growth" forests against lumber companies; the issue revolved around the preservation of the habitats for the Rocky Mountain spotted owl, believed to be an endangered species.

The usual argument runs as follows: Businesses invest capital and create jobs to produce goods and services for a profit. Each company seeks to minimize capital and operating expenses, while maximizing sales and profits. As more environmental restrictions on production are put in place, the cost of producing goes up, which in turn raises the cost of the product and, depending on how vital the product is, reduces its sales. Lost sales reduce employment. Then, when regional or international considerations come into play, manufacturers argue that pollution and jobs are being exported to countries with fewer environmental restrictions, again causing the loss of jobs in the United States—and, not incidentally, not doing the earth a bit of good. Industries with a strong regional aspect to them, such as agriculture, transportation, or recreation, make the argument that restricting the use of a natural resource and reducing pollution in one region gives an "unfair" advantage to other regions that are not so encumbered by regional problems. The automobile industry, for example, argues that increasing the air-pollution-control measures on cars for the sake of Los Angeles's air does no good for the Midwest, where automotive air pollution is not such a problem.

However, newer economic studies have looked at a broader picture. An economics-consulting firm in Washington, D.C., makes a powerful argument that environmental spending sometimes can increase employment. The company, Management Information Services, Inc., did one study of the effects of greater sulfur dioxide restrictions (to reduce the occurrence of acid rain) in the mid-1980s for the State of Ohio. This state has an especially difficult time dealing with acid-rain controls because it has many utility plants that burn coal (the source of the sulfur dioxide), which is itself mined in Ohio. Higher pollution controls would hurt miners and all heavy users of electricity in the state. It so happens that much air-pollution-control equipment is manufactured within the state, and these manufacturers would benefit very directly from increased sales, as would the steel manufacturers, chemical suppliers, and other businesses that supply the air-pollution-control businesses. MIS's analysis showed that if new pollution-control equipment were required nationwide (not just in Ohio) for utility plants, the net benefit to the state would be nearly an additional 6,000 jobs. "In attempting to assess the impact of proposed acid-rain legislation on a state, it is critical to estimate the indirect economic benefits resulting from nationwide compliance, rather than—as is usually the case—simply looking at the likely economic losses to the state," MIS noted.

A rather obvious but often overlooked consequence of environmental spending is that the money isn't simply thrown down a hole and lost to society. Environmental spending goes into the purchase of new equipment, the services of design and construction engineers, the additional employees needed to run or service the equipment, and so on. True, PAC expenditures are a "tax" on manufacturers or American consumers, but those revenues are turned around within the economy and transformed into jobs for other manufacturers. "Indeed, if PABCO [MIS's term for PAC] were a corporation it would rank near the top of the Fortune 500," says MIS.

MIS's studies have been extended to the nation as a whole, independent of regional variations and specific antipollution laws. The analysis is performed first by translating pollution-control programs into economic outputs of all industries in the economy. Such outputs have secondary effects; an increase in fabricated metal production, for example, results in more steel production (a direct effect). But indirectly, it also results in more production of chemicals (which are needed to produce the steel). These relationships are determined by "interindustry inverse equations," according to MIS, and the analysis can be repeated a number of times to derive the total effect of the PAC investments. Economic output data are then compared to a model of the U.S. work force, which is organized according to industry segments.

Based on the 1988 job force and the structure of the U.S. economy in 1988, MIS found that a total of 2.96 million additional jobs were created by PAC (see Figure 2). Within the engineering professions, a full 79,850 new jobs were created (see Figure 3). With results like these coming out of the economic-research community, even *The Wall Street Journal* (whose editorials can reliably be depended on to oppose nearly any new environmental legislation) was compelled to say, in January 1986, "It is interesting to observe, though, how the capitalist system has turned environmentalism into an industry of its own—with its own jobs, its own investments, its own need for markets."

The exact number of jobs, and their types, analyzed by such macroeconomic models are perhaps not as important as the overall implication of the MIS analysis: Money spent on the environment is, at worst, a partial loss to the strength of the U.S. economy. A number of other economic studies have demonstrated that pollution, in and of itself, represents wasted effort on the part of the manufacturer producing it. A company that has fewer pollution problems, generally, is more profitable than a polluting company. Economic studies such as these, from the bastions of business management, are going a long way to change the thinking of companies that heretofore had a reflex reaction against tighter pollution control.

This realization, in turn, leads to a very important implication: Environmental spending is no longer dependent on the good will of the U.S. business community, or the popularity, as measured by public-opinion polls, of environmental causes. Hard-nosed business managers who would have willingly cut employment or shut down factories when business conditions soured are now saying, "This plant pollutes and that costs us lost profits. Fix it!"

Manufacturing, total	1,151,600
Textile and apparel	27,800
Chemical and petroleum products	164,800
Fabricated metal products	60,200
Machinery	133,100
Transportation equipment	81,700
Other manufacturing	684,000
Agriculture, forestry, and fisheries	74,400
Mining	193,300
Construction	111,700
Transportation, communications, and utilities	479,800
Finance, insurance, and real estate	73,100
Services	325,900
Government (federal, state, local)	332,600
TOTAL, all sectors	2,963,400

FIGURE 2   PAC Jobs Created in 1988 (*Source*: Management Information Services, Inc.)

Aeronautical/Aerospace	2,630
Chemical	8,690
Civil	9,060
Electrical	18,200
Industrial	12,130
Mechanical	10,450
Metallurgical	940
Mining	1,440
Petroleum	2,070
Sales	1,210
Engineers, not elsewhere classified	13,030
TOTAL, all engineers	79,850

FIGURE 3   PAC Jobs for Engineers in 1988 (*Source*: Management Information Services, Inc.)

The Rule Book of Environmental Laws

Simply writing new environmental laws, let alone changing manufacturing or land-use policies to adhere to them, is a major task in Washington these days. The revised Clean Air Act, for example, took up nearly as many resources (in terms of legislators' time, committee meetings, lobbying efforts, business and scientific studies, and so on) as the nation's budget took in the fall of 1990. The law, which will cost the U.S. economy some $21.5 billion per year, ultimately was passed in November 1990. Somewhat earlier, in March 1990, *The Wall Street Journal* commented in a front-page story that "the wallop [that] will be delivered by Congress in legislation proposed by President Bush [will] have a wider impact on American business than anything since the 1986 tax-reform law" (p. A1).

Such pieces of legislation are the foundation blocks of the U.S. environmental effort. Green-collar jobs are created with each new law. And, as Figure 4 demonstrates, the number of laws being passed each year is zooming upward. It is an exaggeration, but with a germ of truth, to say that jobs such as a "Clean Air Act engineer" or a "Superfund geologist" are being created. With all this in mind, a review of the major federal laws (and some state laws) is in order. Understanding how these laws are constituted provides a necessary background to the description of environmental businesses in the next chapter.

The Wilderness Act of 1964

Since the late 1800s, when the true worth of the great vistas of the American West became known, there has been a constant tug-of-war between, on the one side, government regulators and private industry, and nature-lovers on the other. The legislative actions that went into Theodore Roosevelt's campaigns for nature preservation became, over the years, interpreted as a federal mandate to "manage" wilderness. All too often, management meant exploitation. The Wilderness Act of 1964 was passed to attempt to regulate the regulators. The law charged the Department of the Interior and the Bureau of Land Management (see? there's that word again) to restrict use of national forests and other federally owned lands. Priorities were to be set over what lands could be used for industrial or agricultural purposes, and wilderness reserves were to be established.

The Wilderness Act was later complemented by the Federal Land Policy and Management (there's that word again) Act of 1976, which

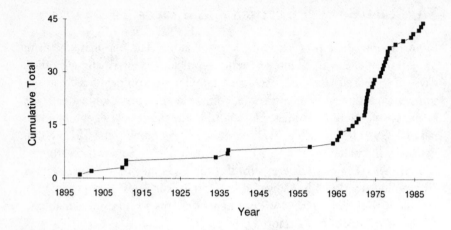

1899 River and Harbors Act (RHA)
1902 Reclamation Act (RA)
1910 Insecticide Act (IA)
1911 Weeks Law (WL)
1934 Taylor Graring Act (TGA)
1937 Flood Control Act (FCA)
1937 Wildlife Restoration Act (WRA)
1958 Fish and Wildlife Coordination
     Act (FWCA)
1964 Wilderness Act (WA)
1965 Solid Waste Disposal Act (SWDA)
1965 Water Resources Planning Act
     (WRPA)
1966 National Historic Preservation Act
     (NHPA)
1968 Wild and Scenic Rivers Act
     (WSRA)
1969 National Environmental Policy Act
     (NEPA)
1970 Clean Air Act (CAA)
1970 Occupational Safety and Health
     Act (OSHA)
1972 Water Pollution Control Act
     (WPCA)
1972 Marine Protection, Research and
     Sanctuaries Act (MPRSA)
1972 Coastal Zone Management Act
     (CZMA)
1972 Home Control Act (HCA)
1972 Federal Insecticide, Fungicide and
     Rodenticide Act (FIFRA)
1972 Parks and Waterways Safety Act
     (PWSA)
1972 Marine Mammal Protection Act
     (MMPA)

1973 Endangered Species Act (ESA)
1974 Deepwater Port Act (DPA)
1974 Safe Drinking Water Act (SDWA)
1974 Energy Supply and Environmental
     Coordination Act (ESECA)
1976 Toxic Substances Control Act
     (TSCA)
1976 Federal Land Policy and Manage-
     ment Act (FLPMA)
1976 Resource Conservation and Re-
     covery Act (RCRA)
1977 Clean Air Act Amendments
     (CAAA)
1977 Clean Water Act (CWA)
1977 Surface Mining Control and Recla-
     mation Act (SMCRA)
1977 Soil and Water Resources Con-
     servation Act (SWRCA)
1978 Endangered Species Act Amend-
     ments (ESAA)
1978 Environmental Education Act
     (EEA)
1980 Comprehensive Environmental
     Response Compensation and
     Liability Act (CERCLA)
1982 Nuclear Waste Policy Act (NWPA)
1984 Resource Conservation and Re-
     covery Act Amendments (RCRAA)
1984 Environmental Programs and
     Assistance Act (EPAA)
1986 Safe Drinking Water Act Amend-
     ments (SDWAA)
1986 Superfund Amendments and Re-
     organization Act (SARA)

FIGURE 4 The Rising Tide of Environmental Legislation (*Source*: Adapted from Yeager, K. E., and Baruch, S. B., "Environmental Issues Affecting Coal Technology." *Annual Reviews of Energy*, 1987, Vol. 12, pp. 471–502)

further classified federal lands among categories of complete wilderness, mixed use, and available for most typical industrial, agricultural, or recreational purposes. Major legal battles sway back and forth over how wilderness lands can be preserved or exploited. In employment terms, these laws created the need for better surveys of existing plant and animal life and a better understanding of geological conditions among the forest lands. The reclamation of already-exploited land also calls for scientists knowledgeable in forestry and biology.

The National Environmental Policy Act of 1969 (NEPA)

This law ushered in the modern era of environmental protection, being enacted as the environmentalist movement of the 1960s was peaking. The emphasis of the time favored wildlife preservation and better management of natural resources, rather than the direct control of pollution. While encouraging "harmony between man and his environment," the law laid out procedures by which an "environmental impact statement" (EIS) was to be produced by any federal agency contemplating significant construction or development projects, such as those conducted by the Army Corps of Engineers or the Bureau of Reclamation. The permit to allow construction to proceed would be predicated on the results of the EIS. Public participation in the EIS review process was also implemented by the law. NEPA eventually became the model for the actions of other federal agencies besides those concerned with construction projects; states followed suit, and soon private industry was obliged to develop EISs as well.

EISs have been a major battleground between proponents and foes of many large projects, as well as evolving technologies. The Trans-Alaska Pipeline, carrying oil from Alaska's North Slope to the south, was held up for years by debates over its EIS. When the Endangered Species Act was passed during the 1970s, an evaluation of the effects of a project of indigenous life became part of an EIS. By demonstrating that there was a rare species of plant or animal life in the vicinity of a project, that project could be delayed or abandoned.

An unknown number of development projects have been forsworn out of doubts that the EIS would pass muster. A handful of other projects, heavily publicized in recent years, have been delayed or stopped altogether by the inability (or unwillingness) of government agencies and private developers to take the requirements of the Endangered Species Act seriously in their EISs. The snail darter, a thought-to-be-rare fish, held up a dam in Tennessee; in New York City, a multibillion-dollar highway construction project in Manhattan was killed by the inability of the agencies involved in pursuing it to account for its effects on fish life in the river bed where dredging would occur. Most recently, a project to build a telescope in Arizona was derailed by fears over its effects on a species of squirrel in the area. Jeremy Rifkin, an antibiotechnology activist in Washington, repeatedly has been able to hold up tests or permits for genetic engineering projects by seeking injunctions against federal agencies for not performing an EIS before allowing these projects to proceed. (Even so, most of Rifkin's suits have been thrown out of court.)

One employment aspect of NEPA and the Endangered Species Act has been to greatly increase the demand for biologists and naturalists to survey the plant and animal life in the region surrounding a project. To follow the intent of the laws, a developer should perform this survey thoroughly. Conversely, an opponent of a project can seek to delay or kill it by performing its own surveys. The process of writing an EIS, too, has become a skill in great demand, involving equal parts of law, biological science, engineering, and public policy making.

The Clean Air Act of 1970 (CAA)

Constructed on earlier laws written in the 1950s and before to regulate smoke (especially around cities), the Clean Air Act of 1970 (CAA) was the first attempt to establish national standards for air quality, and the first to take on nationwide air-pollution sources such as industry smokestacks and automobiles. With this law, too, a structure began to evolve whereby the federal government sets minimum standards for environmental quality, and states, under federal review and approval, develop methods to regulate pollution sources within their borders. Air pollution, in contrast to other media such as solid waste, is automatically a national issue; a molecule of sulfur dioxide ejected into the air in New Mexico may ultimately become a tree-killing pollutant in South Carolina. CAA was revised in 1977 and again in 1990.

CAA set "national ambient air quality standards" (NAAQS) for specific pollutants, such as sulfur dioxide, nitrous oxide, carbon monoxide, and ozone. State implementation plans (SIPs) were written to carry out the actual enforcement. The regulations originally made a distinction between existing sources of pollution and new ones (i.e., a new factory or utility plant); the latter had to meet "new source performance standards" (NSPS). Another hallmark of CAA was the first prominent use of recommended technologies, such as scrubbers for sulfur oxides. In general, industry is much more comfortable with standards that specify a level of performance, while letting the polluter itself decide how to meet that standard. A later revision of CAA regulations, put forth in the early 1980s, established a "bubble" policy for factories or other stationary sources of air pollution. In essence, an imaginary bubble is put around a plant, and the total amount of pollution emitted is measured. The plant owner then can choose what pollution sources to go after first in order to reduce overall levels of pollution.

CAA and its revisions have helped turn around the growth rates of pollution emissions in a dramatic fashion, having reduced ambient levels of some gases by 30 percent or more since 1970—over a period of time when U.S. economic activity has just about doubled and population has risen by a quarter. However, as alluded to in other parts of this book, this "success" is often measured against a criterion of zero emissions. In actual terms, zero emissions is impossible; the mere act of breathing emits carbon dioxide, a gas that affects the ozone layer of the Earth. The perception of the rate of reduction in emissions can be measured against the force with which the American people and their elected representatives push for speedier reductions.

A number of cities have consistently been unable to meet CAA standards. At first, Congress allowed moratoriums on enforcement, but with the latest revision of CAA, more specific regulations must be met. Now, in the 1990s, cities will be chartered to experiment with alternative means of meeting NAAQS, such as the use of electrically powered cars or restrictions on the use of automobiles entering city limits.

A final element of CAA, in its latest incarnation, is also one of the most innovative: marketable pollution rights. In essence, this provision says that a molecule of pollution in one region of the country can be just as much of a problem as a molecule in another part (which is truest in cases where pollution is a national problem; other gases, such as ozone around cities, are not spread throughout the countryside). Company A, having installed new equipment to reduce emissions, also acquires the right to market this foregone pollution to other companies which, instead of also installing equipment, will purchase that "right" to emit. If the system works as it is intended, the marketplace will decide which companies have the most expensive problems to solve (in terms of the cost of preventing pollution). These companies will purchase pollution rights from other companies, whose emissions are less expensive to prevent. This purchase price will help defray the cost of the equipment installation. In this way, pollution for the nation as a whole will be reduced in the most economically sensible fashion. As of this writing, the details of how the program will function were still being worked out. But one can readily see that the skills of a stockbroker or financial analyst are as meaningful to running this program as the skills of an engineer are to building a pollution-control system.

The Clean Water Act of 1972

The Clean Water Act that was passed in 1972 built on one of the oldest environmental laws, the Rivers and Harbors Act of 1899; the latter was also updated by the Federal Water Pollution Control Act of 1948. As with air pollutants, water pollutants affect many regions once they are dispersed into a river. The key acronym of CWA's regulations is NPDES, which stands for National Pollutant Discharge Elimination System. On a case-by-case basis, and with sometimes highly specific (and theoretical) considerations, a private company or public agency that seeks to discharge contaminated water must first receive a permit from the state or federal government. Such pollution is called "point source" pollution, meaning that it comes from an identifiable source. One will often hear about a project being held up because its NPDES permit has not been obtained.

CWA created a very expensive program when it was passed—the Construction Grants program. Originally, it had funding that allowed the federal government to pay up to 75 percent of the cost of building a municipal waste water treatment plant; the city or state would kick in the other 25 percent. Billions were spent in this program during the 1970s; when Ronald Reagan came to office, the amount of money allocated for it was dramatically reduced, and the ratio of federal contributions was reduced. Even so, the overall national effort for waste water treatment has been described as the second largest public-works program in U.S. history (behind the national highway program).

Such municipal plants (often called "POTWs," for publicly owned treatment works) play a central role when it comes to the nation's water quality. POTWs treat sewage, usually by speeding up the natural biological degradation of organic matter in waste water. Many manufacturers tie their water-disposal system into the POTW and pay a fee to allow it to be treated. A sludge is produced, which itself is a waste that must be disposed of. Many cities along coastal regions simply barge or pump the sludge out to sea; this is the practice in Boston Harbor that led, in part, to the downfall of Michael Dukakis as a presidential candidate in 1988. As of the end of 1991, this practice will be restricted, with a fee being paid by the municipality to continue it.

A dizzying array of acronyms have been created to specify what constitutes appropriate treatment, including best practicable technology (BPT);

best available technology (BAT), best conventional technology (BCT), and others. The reason for all these qualifications is that it is exceedingly difficult to control the quality of water entering a river or lake, regardless of what cities or industries do. A heavy rainfall, for example, quickly overwhelms the processing capacity of a POTW, with the result that raw sewage (much diluted by the rainfall) is swept out. In addition, a substantial amount of water pollution comes from "nonpoint" sources: the runoff from city streets or agricultural farmlands. Finally, because such forms of pollution as sewage are degraded by natural processes, there will be areas where water pollution can be discharged into a river or lake with very little measurable damage; the same amount of pollution discharged into another body of water would quickly kill most fish life in it. There has been a constant tug-of-war between industrial firms, which have the option of treating their own waste water or passing it onto a POTW; cities, whose budgets are increasingly consumed by waste water treatment; the federal government, which is called on both to set high pollution standards and to foot the bill for meeting them; and wilderness preservationists, who seek to drastically reduce all such pollution.

A further complication of CWA is the fact that much of the world's water lies underground—groundwater. This is the source of drinking water for half of the nation and has increasingly been polluted by garbage dumps, injection wells, underground storage tanks, and other sources. Groundwater figures largely in other environmental legislation, including the Safe Drinking Water Act and the Resource Conservation and Recovery Act (see below).

The Safe Drinking Water Act (SDWA)

Passed in 1974, SDWA is designed to regulate the 60,000 water treatment systems in the nation. The main problem with drinking water drawn from surface waters is that it is polluted with bacteria-bearing sewage or contains microorganisms that can harm human health. In most cases, a dose of chlorine kills these microbes, and, after filtering, the water is fit to drink. But this covers only half the nation; the other half depends on water drawn from wells (groundwater). Here, the insidious nature of sloppy waste disposal or storage becomes apparent. Chemicals that were dumped in a pit and covered over slowly leach out, creating a plume of contamination. When it reaches a well, that well becomes unusable for decades to come. SDWA of itself hasn't spawned a major construction or regulatory program, but has served to set contamination levels for types of chemicals or microbes in drinking water. These standards, in turn, have become an important regulatory limit for other laws pertaining to water quality. Amendments to the law in 1986 added to the list of chemicals whose concentration in water are to be regulated.

The Resource Conservation and Recovery Act of 1976 (RCRA)

The oddly named RCRA law, first passed in 1976 and reauthorized in 1984, has had one of the most dramatic effects on industry and commercial practices of any environmental legislation. It is oddly named because, while indicating that it pertains to "resources" and "conservation," to this date it is primarily about a very different thing: hazardous wastes.

What exactly are hazardous wastes? By EPA definition, they are the waste byproducts of manufacturing that are ignitable (flammable), corrosive (strong acids or bases), reactive (unstable chemicals that explode or generate other toxic substances), or toxic to human health. Certain types of chemicals are listed as hazardous by default; if they are present in a waste, that waste is hazardous (R.C. Fortuna and D.J. Lennett, *Hazardous Waste Regulation*, p. 40). By applying these criteria to literally thousands of streams of byproducts coming from human activity (either industry or society at large), more than 200 million tons of hazardous wastes per year came under regulation. With broader definitions established in the 1984 amendments, the total volume has risen to 360 million tons per year. To begin with, this is a fantastic amount of materials; if each citizen were personally responsible for his or her proportionate share, it would amount to nearly one-and-a-half tons of oily sludge, gooey slops, caustic liquids, or lumps of junk producing smoldering, lung-searing fumes.

The second problem is that RCRA establishes very selective criteria for what can be done with these wastes. Today, there are only a few dozen landfills where they can be dumped into the ground in the country. Transporting the wastes from where they are generated to where they can be dumped is a further complication; in some locales, residents have formed human chains across roadways to prevent the passage of the wastes. A significant volume of hazardous wastes is actually waste water, which can be treated in facilities similar to those that treat sewage, but the cost of doing so is not inconsequential, and the sludge that remains after processing is still hazardous. (Much of the water, however, has been removed.)

The third problem hazardous waste generators face is meeting the regulatory requirements of RCRA. If the proper permits have not been received for generating the waste in the first place, for storing it on site properly, and for transferring it to a properly permitted waste hauler, who in turn brings it to a permitted disposal or treatment facility, the liabilities are severe. This concept has been called "cradle-to-grave" coverage of hazardous wastes. Corporate executives are regularly going to jail for not facing up to the requirements of the law; multimillion-dollar fines have been paid. If one link in the chain from the waste generator to the waste disposer is broken, the liabilities snake right back up the chain to the generator. Some EPA officials responsible for enforcing RCRA laws carry guns.

Before you think that these liabilities are the just consequences of polluting industrialists, you should realize that we all are generators of hazardous wastes. Thrown an empty paint can into the trash? That's a hazardous waste. Dumped dirty cleaning solvent down the drain or burned oily rags in the back yard? That's improper disposal of hazardous waste. The only reason EPA hasn't been knocking on your door is that the law sets a limit of 100 pounds per month as the minimum that defines a hazardous waste generator. But even that limit, set by the 1984 amendments to the act, brings in many auto repair shops, dry cleaners, university laboratories, and other Main Street enterprises.

The biggest waste generators, of course, have the biggest liabilities and have devoted the most resources to meeting RCRA standards. This is primarily the chemical industry, metals and petroleum refiners, and fabricated-metal producers. Some of these companies, having become expert at analyzing wastes and formulating disposal solutions, have now gone into the business of providing consulting services. A multibillion-dollar industry to burn, concentrate, and dispose of hazardous wastes has come into being over the past decade. "RCRA is one of the most controversial, emotional and political of any federal regulations," says William Lorenz, a Concord, New Hampshire, business consultant to the pollution-control industry.

RCRA was also responsible for creating a syndrome, now applicable to great swaths of economic activity, called NIMBY—not in my back yard. NIMBY arose when industries sought, while fulfilling every statutory requirement of RCRA for building a hazardous waste treatment or disposal facility, to find sites for the facilities. No matter where they went, nearby citizens fought back, holding marches, bringing lawsuits, and excoriating the local and state officials who allowed the business even to consider bringing such a facility to their neighborhood. State officials, recognizing that unless some means of creating disposal capacity for industry were created, those industries would be forced to shut down and leave the state, have also been rebuffed. The NIMBY syndrome has since come to be applied to new factories of any type (whether or not they generate hazardous wastes), real estate developments, and public works—almost any concentrated form of economic activity. State regulators themselves have gotten into the NIMBY act: Some states have instituted bans against receiving wastes from other states, not wanting to be the dumping ground for someone else's waste.

For all the grief RCRA is causing industry, some positive benefits are beginning to appear. Government regulators and industry analysts now know more clearly where wastes are coming from and where they are going; emergency situations, when a community suddenly realizes that its very life is at stake because of the appearance of hazardous waste, are becoming rarer. In addition, although the recorded volume of wastes has not fallen, their nature is changing. Companies are becoming more savvy about handling raw materials, recycling materials back into the production process, or basically plugging leaks in processes where materials had been escaping. The operators of hazardous waste disposal facilities report that the types of wastes they are handling are becoming more concentrated, with lower volumes. (Again, it bears emphasizing that the total national volume of hazardous waste has been rising because more categories of wastes are being regulated.) Where possible, hazardous waste generators are processing their wastes at their own sites, which is generally preferable to endangering the populace by transporting wastes through public roads and transferring a waste problem to someone else.

Most significantly, many companies are beginning to rethink their entire manufacturing processes and lines of business. Production techniques are now being tested for their environmental characteristics before a factory is built. "Our company has consciously avoided several businesses over the past decade because of the difficulties in disposing of the hazardous wastes that would be generated," avers one chemical company executive. For existing processes, manufacturers are actively pursuing a philosophy called waste minimization, whereby a process is reevaluated and changed to reduce or eliminate waste production. Waste minimization finally, 15 years after passage of RCRA, gets back to its original intent: the conservation of resources and their recovery where it is possible.

RCRA has generated thousands of green-collar jobs in industry, federal and state government, consulting, and research. Much of the responsibility for enforcing it falls on the shoulders of state regulators, so growth there has been especially significant. The tens of billions of dollars spent either to dispose of wastes or to reengineer processes to eliminate wastes makes ready markets for the experts. In addition, the hazardous waste disposal industry, which had been the ugly stepchild of municipal sanitation and garbage hauling, has now come into its own as an industry. Chemical Waste

Management, Inc., a subsidiary of Waste Management, Inc. (Oak Brook, Illinois), is now nearly a $1 billion a year company, employing about 4,500, and is probably the largest such private enterprise in the world. Its revenues rose 27 percent in 1989.

The Toxic Substances Control Act of 1976 (TSCA)

TSCA, usually spoken as "toss-ka," is concerned with the new production of materials that could have serious environmental effects. One of the key provisions is the requirement that manufacturers submit a "Premanufacturing Notification" (usually referred to as a PMN) when the scale-up of production of some material is contemplated. In essence, the PMN gives EPA a chance to review the potential environmental consequences of a chemical before it actually goes into production. A key element of winning EPA permission to manufacture is the submission of toxicological and other data. PMNs are a constant worry among pharmaceutical companies, which must first meet EPA's TSCA requirements before then running the gauntlet of approvals from the U.S. Food and Drug Administration. (FDA not only requires that a medication be harmless when dispensed properly; it also requires that the medication do what it is supposed to.) TSCA is a bit of an EPA/government power play; in certain contexts, it gives EPA priority over other agencies or over state regulations.

The Comprehensive Environmental Response, Compensation, and Liability Act (CERCLA, or "Superfund")

CERCLA, passed in 1980, is the offspring of Love Canal (see Chapter 2). That unregulated waste dump, covered over years before, had a school built on top of it and was eventually discovered to be leaking toxic materials through the groundwater in the surrounding neighborhood. President Jimmy Carter declared the site to be a national emergency and, after much litigation, the state purchased the homes in the vicinity and essentially condemned and demolished much of the neighborhood.

CERCLA was legislated to cover several gaping holes in environmental law and to make amends for the mistakes of history. There are a couple terms that usually come up when CERCLA is discussed. One is "Superfund," another name for CERCLA, which refers to the provisions of the law that set up a billion-dollar fund to conduct cleanups of old, abandoned dumps. Another is SARA, which stands for Superfund Amendments and Reauthorization Act, which was passed to update Superfund in 1986. SARA significantly altered and expanded the Superfund law.

Superfund has also been called, venomously, the Environmental Lawyer's Full-Time Employment Act because of the intense amount of litigation it has generated. "Driven by the needs of industry, seemingly thousands of hours of lawyer time are now spent in . . . Superfund-related activities," wrote Ann Powers, director of the Chesapeake Bay Foundation (Charles Openchowski, A *Guide to Environmental Law in Washington, D.C.*, p. 198).

By and large, Superfund addresses the same problem as RCRA: the disposal of hazardous wastes throughout the country. The major difference is that it addresses wastes that have already been dumped or contamination that has already occurred, rather than ongoing production or contamination. The most expensive element, in terms of government revenues, is a tax on the production of various industrial chemicals and materials, which go into a trust fund to pay for the cleanup of abandoned dump sites for which no "potentially responsible party" (PRP) can be found. When a PRP can be identified—sometimes through attempting to read shipping labels on steel drums that were buried years ago—EPA lawyers go after the PRP to win monies for completing a cleanup. Many companies, having complied fully with what they felt were responsible disposal practices of the 1950s or 1960s, now find themselves paying millions of dollars to clean up these old dumps.

With EPA oversight, Superfund contractors perform an RI/FS (remedial investigation/feasibility study) of a suspected dump site. If significant contamination is found, the site becomes a candidate for inclusion on the National Priorities List (NPL) that EPA has compiled. There follow a series of studies of the possible remediation methods for cleaning up or at least stabilizing the site (to prevent the spread of contamination) and the awarding of contracts to carry out the actual cleanup work. At this point, the question "how clean is clean?" takes on momentous impact. Each project is extremely site-specific, both in terms of the types of wastes present and the geological or habitation factors at the site. Site remediations can cost tens of millions of dollars. The NPL has some 1,200 listings, and there are thousands more sites that may eventually be added to it.

When the original Superfund was passed, it was written with Love Canal very much on the minds of legislators. They wanted to make sure that persons in the vicinity of a dump site could gain compensation for the environmental degradation their property suffered. An element of enforcing these payments was the inclusion of the legal term "joint and several liability" in the law, which specifies that a PRP can be held accountable for the actions of, say, the waste dump manager, even when the PRP had no control over what that manager was doing. An analogy would be that you yourself could be arrested for bank robbery if you were a customer in the bank while a robbery was occurring. This is, of course, an extremely strong legal club, but legislators at the time felt it was essential to clear away endless legal debates over who was responsible for contamination at an old dump. Again, the need for legal services expanded greatly when this rule was legislated.

Another element of Superfund was the formalization of an emergency-response resource when environmental pollution occurs, usually as the result of an accident or natural disaster. EPA now employs contractors for each of ten regions in the country, which are supposed to be ready to jump into action when, say, a tank truck carrying toxic chemicals overturns on a highway. This emergency-response mechanism meshes with a variety of civil-defense or insurance-industry measures to respond to accidents and disasters and demonstrates how such accident specialists are now part of the green-collar work force.

In 1984, an industrial accident occurred in India that had great repercussions in the United States. This was the Bhopal chemical spill, which caused thousands of deaths when a Union Carbide chemical plant released a cloud of toxic methyl isocyanate gas in the middle of the night. Superfund was up for reauthorization at the time, and legislators quickly pounced on it to expand Superfund's authority. The reauthorization, SARA, now includes what has come to be called "community right-to-know" rules, which stipulate that toxic chemical manufacturers need to coordinate emergency-response measures with local authorities (police, fire, etc.) and with the federal-level emergency-response program. Manufacturers are also required to be more forthcoming with information about dangerous chemicals for their workers. This is carried out by the composition and dissemination of a "material safety data sheet" for workers, local community representatives, and customers who buy the product. In essence, these right-to-know provisions fill the gap in the RCRA cradle-to-grave scheme between when a chemical is made and when it is used. However, they apply to all chemicals or hazardous materials, not just wastes.

Manufacturers were extremely loath to commit to community right-to-know, because often this was considered to be proprietary information that a competitor could use to discover trade secrets about how a product is made. The more socially conscious manufacturers (or the far-sighted ones who wanted to be prepared for accidents) had made sure that enough staff people were on hand around the clock at a plant to know how to deal with a spill, accident, or disaster. Now, however, everybody had to come clean.

This communication channel has greatly increased the need for green-collar communicators, ranging from the people who write material safety data sheets, to those who meet with community leaders or the community at large to assure them that these dangerous materials are being handled responsibly. Companies that are relatively uncooperative with local communities soon find themselves in court, paying for expensive lawyers' time when many of the problems could have been addressed more effectively by a good communications program.

Dump-site remediation is mostly an exercise in civil and geological engineering. Sites have to be examined, tests made of samples, and surveys of rock and earth strata and groundwater flows conducted. When the site is to be cleaned up, risk assessments are performed on the various techniques to be used; the cleanup of a dump site could itself cause a severe hazardous material release. Many times (all too often, in the view of many environmental engineers and activists) the most cost-effective practice has been to pick up the contaminated dirt and pump out the contaminated groundwater and cart both of them to a regulated, licensed hazardous waste landfill. Since more than a few legal landfills from the past are now Superfund sites, there are some who worry that today's Superfund cleanup will become tomorrow's new Superfund site, repeating the cycle.

Scientists and engineers have been busy at devising new cleanup techniques, ranging from incineration of soil in high-temperature furnaces, to the use of ultraviolet light to decontaminate water, to the activation of indigenous or scientifically selected bacteria which can degrade certain compounds *in situ*. EPA, recognizing the need for new technology, runs a program called SITE (Superfund Innovative Technology Evaluation), in which a new technology can be tested, under carefully controlled conditions, at a Superfund site.

Superfund, more than any other law, has brought the insurance industry into the green-collar work force. Many PRPs, having dusted off decades-old insurance policies, recognize that their liability may be covered by the policy, depending on how it was written, and notwithstanding the knowledge that no one realized the consequences of dumping hazardous wastes in the ground. Now, manufacturers who have been hauled into court as PRPs are turning around and suing their insurers. More lawyers' time. The insurance industry is important for two other reasons as well. More than anywhere else, the expertise to perform risk assessments resides in the insurance industry. Superfund requires such assessments for ranking sites, for evaluating health risks, and for selecting preferred remediation technologies. A second type of expertise, available both in industry and among insurers, is "loss control"—the prevention of accidents by establishing wiser manufacturing practices and by making preparations for accidents when they do occur. For many years, industrial insurers have been accustomed to examining a manufacturer's facilities and training prior to writing a policy for it; if the manufacturer puts recommendations into practice, a better rate of insurance is provided. Now, with the environmental liabilities of a spill, an improper disposal of waste, or even the unnecessary use of a hazardous material representing a potentially huge financial risk, manufacturers are turning to the insurance industry for advice on becoming more environmentally conscientious. This is another case of corporations becoming environmentally good citizens purely for profit motives.

Finally, relations between insurers and the insured have become testy, to say the least, as a result of Superfund. A significant number of manufacturers, unable to win policies from insurers that they are satisfied with, are seeking self-insurance. (When you stop to think about it, your own retirement plan is a form of insurance, assuring you of maintaining a livable income when a job stops due to retirement. In this same spirit, a company can set aside funds in anticipation of a future accident or environmental liability.) To self-insure, a company needs access to the same underwriting and investing expertise that insurance companies employ.

State and Other Laws

There are a great number of laws pertaining to wildlife, fisheries, coastal areas, and other natural resources that were not described in detail here. These include:

- National Wild and Scenic Rivers Act of 1968
- The Marine Protection, Research and Sanctuaries Act of 1972
- The Federal Land Policy and Management Act of 1976
- The Surface Mining Control and Reclamation Act of 1977

In reading the preceding pages, you will have observed that next to nothing is said about nuclear wastes, which were mentioned in the previous section as a hundred-billion-dollar-plus problem. There are a variety of reasons for this; most nonnuclear laws were written to exclude radioactive wastes, because they present an entirely different set of risks and liabilities. It is also easy to see—now, in hindsight—that the federal government, the sole manufacturer of nuclear weapons, was loath to write rules regulating itself. Nevertheless, the processing facilities managed by the U.S. Department of Energy (DOE) were inherited from the Energy Research and Development Commission in the 1970s, and before that, the Atomic Energy Commission. The law that set this commission into being was signed in 1954, and some of the disposal practices now being sought for radioactive wastes will be in compliance with that law. DOE is now in the process of meeting the standards of RCRA with nonradioactive wastes, at least, and some disposal sites for some of the wastes exist already.

In addition, there are a great number of state-level laws that often provide a substantial impetus to environmentally related activities. Most states have set up quasi-governmental commissions (which allow for public participation) to devise methods for siting waste landfills and for allocating natural resources among industry, agriculture, recreation, and wilderness preservation. The latest batch of laws set up recycling programs; many of these, however, are city ordinances, rather than state ones. Western states have elaborate programs for allocating water resources; coastal states have programs for beaches, estuaries, and river mouths.

California remains at the forefront of environmental legislation, both because of its high-minded citizenry and because of its dramatic environmental problems. The City of Los Angeles has one of the worst air-pollution problems in the country because of the coastal valley it lies in, which prevents a rapid exchange of dirty air. The entire state has suffered through five straight years of drought, beginning in the mid-1980s, and there is no end in sight as of today. Finally, the enormous agricultural business in the state attunes people to such issues as land use, pesticides, food purity, and wilderness conservation.

CHAPTER FOUR

Environmental Business II

It has been said about my profession—architecture—that it creates the symbols that define a culture. Five hundred years ago the dominant symbol was the medieval cathedral. One hundred years ago, in the United States, it was the capital dome of a government building. In this century, it has been the corporate high rise tower. In coming years, I expect it to be structures that restore the environment, not destroy it. A new shift is happening, today.

—Robert Berkibile, AIA, Chair of the Environmental Committee of the American Institute of Architects

The previous chapter outlined the size of the environmental business in the United States today and what laws drive it forward. This chapter looks at the players: the government agencies, the manufacturers, the marketers and recreation providers, the consultants, the business services, and the nonprofits. Green-collar workers occupy key positions throughout these organizations.

This chapter will review where green-collar workers are employed and what types of responsibilities they have. It is organized according to industry or organization, not type of profession. You'll have to read the whole chapter to get a sense of, for example, where marine biologists work. More details on employment opportunities will be covered in the next chapters, which are organized according to basic professions. Finally, the order in which each of the groups of organizations is written up is a reflection of the approximate rank (in terms of numbers of jobs) of each, starting with private industry and ending with nonprofits. The order they are ranked in is as follows:

1. Private industry
   a. Pollution generators

     b. "Green" marketers

     c. Recyclers

     d. Disposers

2. Government

     a. Federal

     b. State

     c. Local

3. Business services

     a. Science, engineering, and architecture consulting

     b. Legal services, insurance and risk management, and market research

     c. Communications, publishing, and advertising

4. Travel, tourism, and recreation
5. Teaching, academic and professional
6. Nonprofit organizations

Private Industry

Pollution Generators

Industrialists can argue, with some justification, that they are only a part of society's pollution problems. Individuals produce hazardous wastes, destroy wildernesses, and pour air pollutants from cars and homes. Still, no one questions the wisdom of going after the products that individuals use to produce pollution (such as the lead that used to be in gasoline) rather than after the individual polluter. It is so much more economical, and efficient, to try to cut pollution at the source where polluting products are made, rather than among the end users, who may or may not dispose of such wastes or use such products in an environmentally sound way.

    Thus, it is no surprise that blue-chip corporations regularly get headlines when their activities are the cause of an environmental assault, such as the Exxon Valdez oil spill in early 1989. Big companies tend to be older, so the faulty environmental practices of the past can be laid at their doorsteps, and, generally, they have the "deep pockets" to fund such cleanups. By the same token, big companies manufacture big volumes of products, and

Table 1 *New Capital Business Expenditures for Pollution Abatement*, 1988

Industry	Expenditure (Billion $)	Total Capital Outlay, %
Petroleum	1.65	8.7
Chemical	1.26	7.1
Primary metals	0.91	9.2
Paper	0.72	6.3
Transportation equipment	0.50	3.1
Food and beverage	0.32	2.6
Electrical machinery	0.20	1.2
Machinery except electrical	0.18	1.2
ALL MFG.	6.51	4.1
Public utilities	1.75	3.8

changes they make can shake up whole industries. Du Pont Co., for example, plans to spend upwards of $1 billion to change over its production of Freon chlorofluorocarbon refrigerants to "Suva" refrigerants that do not contain chlorine (see Chapter 7). One billion dollars is not an inconsequential sum by any measure; at the same time, it is also true that some 25 percent of worldwide production of CFCs comes from Du Pont.

According to UNEP data, out of a total of nearly 900 million tons of wastes produced by American society in 1985, 78 percent, or 691 million tons, was produced by industry. This includes all wastes, not just hazardous wastes (A. Hammond, *World Resources 1990–91* p. 325). A look at who in the business community has the most at risk from pollution can be garnered from data provided by the U.S. Bureau of Economic Analysis, shown in Table 1, which details the money committed by business to pollution abatement.

Note a key implication of Table 1: While the average environmental spending of all manufacturing industries in 1988 was 4.1 percent, it was up to double that percentage for the metals, petroleum, chemicals, and paper industries and nearly that much for public utilities. These, then, are the industries with the biggest environmental responsibilities. By and large, they correspond to the traditionally named "smokestack" industries. Following is a rundown of the specific problems each faces.

Primary Metals. "Primary" metals are those formed into raw slabs, sheets, lumps, or other configurations from mineral ores or scrap metal. Smelting—a process where the ore is subjected to heat and chemicals—is the main production technique; it produces heavy amounts of sulfurous and other air pollutants and waste waters contaminated with metals and organic chemicals. When the metal is a dangerous one, such as lead or cadmium, this waste water must be treated carefully.

The steel industry, by far the largest primary metals manufacturer, also depends heavily on coal for making steel from iron. Coal is cooked into "coke" (a purified form of carbon) in special furnaces, and these, too, produce large amounts of air and water pollution. Coal contains large numbers of toxic chemicals, and so the waste waters from the production processes must be cleaned, to a point, in on-site water treatment plants.

Metal refining plants are enormous affairs, with ten-story towers and long buildings all over the plant site. Most refining processes require huge amounts of energy, and so many of the same pollution problems of utility plants exist at these refineries: smoky dusts and slags, sulfur, and the like. This high degree of energy "intensity" (i.e., the amount of energy consumed per pound of product) makes recycling a compelling alternative path. Indeed, up to 60 percent of some metals are produced from recycled garbage. Aluminum is the single most successfully recycled metal, although lead and copper are not far behind. Conversely, many metals companies are integrated with the mines that produce the raw ore for refining, and these mines damage the environment through releasing contaminated water and unsightly piles of slag or rubble.

Petroleum. In the aftermath of the Valdez oil spill, who isn't aware of the potential or actual harm caused by oil drillers? Oil drilling affects the environment in several ways. The wells today tend to be drilled in wilderness areas, which then opens those wildernesses up to abuse when roads are built. Oil drilling requires large amounts of water and chemicals; these contaminated products are sometimes dumped on site and sometimes pumped back underground with the expectation that the pollutants will remain immobile.

Once the petroleum is drawn from the ground, it must be refined, and these refineries produce some of the same wastes as metal refineries: dusts, contaminated waters, sludges, and slags. In addition, because crude oil is the source of many petrochemicals, these sometimes toxic compounds can leak out of process equipment and into the air and water. Many old refineries now have thousands of gallons of oil under their sites, the result of years of spills and leaks.

Chemicals. Chemicals—most often, petrochemicals derived from oil or natural gas—are the most complex of pollution problems. There are some 70,000 industrial chemicals in common use, and each of them has special handling or exposure requirements. The chemical industry is the single largest source of hazardous wastes, and since, at sites like Love Canal all around the country, chemical producers of past decades thought that once a chemical was in the ground, it effortlessly would be degraded to environmentally safe forms, there are now thousands of potential or actual Superfund sites.

Organic chemicals have varying toxicity effects on humans; they can cause cancer; they can make one sterile or cause birth defects; and they can injure lungs, skin, or the nervous system. Therefore, plant-safety procedures are elaborate, and the chemical industry is probably at the forefront of developing community right-to-know functions. In addition to having hazardous wastes that are broadly controlled under RCRA guidelines, there are specific regulations for pesticides, pharmaceuticals, flammable chemicals, and for materials transportation.

The petrochemical portion of the chemical industry is responsible for making the raw materials that go into plastics (and many of them are major plastics manufacturers as well). Plastics have been in the spotlight for years now, somewhat unfairly, as the most excessive example of unnatural products that displace natural ones, while causing litter and garbage problems. There are several moves afoot to recycle plastics, especially the more common ones. Soda bottles, for example, are now being recycled, for the first time, back into soda bottles by the leading suppliers. Plastics are 7 percent of the content of a typical municipal garbage dump on a weight basis and about 20 percent on a volume basis.

Paper. Besides the power utilities, paper companies are the largest industrial consumers of water, and they tend to use pure fresh waters in heavily forested regions of the country. Not coincidentally, those forests are the source of their raw materials. While paper is eminently a "natural" product, its manufacture is not. Wood chips are soaked in caustic or acidic chemicals and cooked, causing portions to dissolve, leaving the valuable pulp. This dissolved "liquor" has many different chemicals in it, ranging from turpentine and pine oils that are used in everyday consumer products, to compounds called furans, which are the precursors of dioxin. Dioxin, a highly toxic material, is produced when the furans are exposed to chlorine in the pulp-making process.

Thus, the paper industry has two major knocks against it: It consumes forests and it contaminates water, often from streams or lakes that fishers and other wildlife devotees cherish. One of the hallmarks of certain types of pulp mills is a characteristic rotten-egg smell (which comes from hydrogen sulfide gas). This noxious chemical doesn't represent a major environmental worry in the concentrations produced by pulp mills, but its unmistakable presence is a significant problem for paper makers.

Paper also has the distinction of being the single largest component of municipal garbage, constituting 36 percent by weight. Paper makers have made great strides in recycling used paper, although the low value of the paper on a weight basis makes it somewhat difficult to justify as a recyclable. In the late 1980s, as environmental interest rose, the sudden surge in paper being offered for recycling dropped the bottom out of the market. Whereas before recyclers could count on revenues by offering the paper back to the industry, they were then confronting a situation where they had to pay to have the paper taken back.

Public Utilities. Public utilities make power, which when fed into an international grid of wires provides the electricity that keeps the lights on. Each step of this process has environmental risks.

First, utilities serve to convert some other source of energy into electricity. This includes using the flow of rivers, which then calls for dams that wreck the natural beauty of some rivers. Or it can be by burning coal, which produces vast amounts of sulfur pollutants, waste ash, and sludge, as well as carbon dioxide, which is now being fingered as the prime contributor to global warming through the greenhouse effect. Or worst of all, in many people's views, it can involve the use of nuclear energy with the concomitant production of high-level fuel wastes. When the power is transmitted over long distances, rights of way are blazed through woods and farmlands to carry the 50,000 volt electricity. And there is increasing evidence, still preliminary at this point, that the power transformers that "step down" this high-voltage current into forms usable by homes and businesses can cause ill health or birth defects.

No one, however, envisions a future without electricity. And while the nuclear energy industry is roundly criticized for its past mistakes or obfuscations, there is a realization that it produces power without vast heaps of coal, ash, and sludge from air-pollution controls and that it does not add to global warming. Other countries, notably Japan and France, have adopted nuclear power to the point where the majority of their electricity comes from it. (These countries have been boxed in, to a point, by the lack of fossil fuels within their borders.) If they can deal with it, why can't we?

There are a great number of alternative technologies being experimented with by the utility industry and the federal government. These include garbage-to-energy, heat from underground sources (geothermal energy), combined electricity and steam production (cogeneration) fueled primarily by cleaner burning natural gas, and solar energy, either in the form of direct conversion of sunlight to electricity or through the intermediaries of sunlight-to-steam or wind energy. All these technologies, however, work in special cases or are limited by high costs. Many of them, too, have their own pollution problems.

By the mid-1980s, utility managers realized that the cheapest kilowatt of power they could buy was the one that their customer didn't use. By helping customers lower electrical demand, utilities could postpone the construction of new power plants. In this analysis, they were helped dramatically by the Environmental Defense Fund. The collaboration represented something of a breakthrough for both industry and environmentalists, getting both of them to work together to solve a pressing problem.

Green-Collar Industrialists. The key professional in all these examples of heavy industry is the engineer, who is usually hired to design, build, and run the factories. Many of the mid- to upper-level managers of industrial concerns also have an engineering background. Over the years, specific engineering disciplines have developed for the industries that need them most: metallurgical engineering for metals refining, petroleum engineering for oil production, and chemical engineering for the chemical industry. The paper industry depends on those trained in chemistry as well as in engineering, and the utility industry depends on a subset of the large electrical engineering community, utility or power engineers.

It is noteworthy that many of these industries have had recycling programs of one sort or another for years. These programs were started up mainly for economic reasons—a cheap source of wood fiber or lower energy costs for metals. Although there are exceptions (the aluminum companies come to mind, with their widely advertised recycling campaigns), most of these manufacturers did not develop their own subsidiaries to purchase recycled goods. Rather, they depended on the scrap and waste-hauling entrepreneurs—in a word, the junk men. In earlier times, this made some sense because the business skills needed to succeed at refining or manufacturing were quite different from those needed to get scrap out of the waste stream. This arrangement is changing dramatically, as both the manufacturers and the scrap dealers uplift the quality and scope of recycling operations. The technologies that most recyclers use today depend in large part on the systems that were developed in the past for in-house recycling.

Heavy manufacturing is hard, dangerous work. As a result, over time these industries have built up a fairly diverse range of safety and loss-control operations. As environmental issues came to the fore, responsibility for that

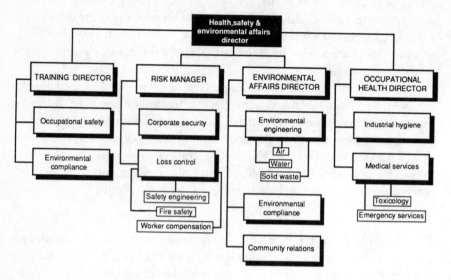

FIGURE 5   Typical Industrial Organization Chart

concern was worked into the existing structure. In fact, business managers soon realized that "environmental" was the one word that unified all these concerns, and thus, "environmental management" has taken over as the operative concept in the corporate chain of command. A typical organizational chart for a large, heavy-industry manufacturer is shown in Figure 5. Note that training is an integral component of the environmental function.

Some companies keep a sturdy division between environmental issues that exist within the factory gates (workplace safety, worker exposure to chemicals, worker training) and those that exist outside the gate (air, water, and solid waste emissions, governmental regulatory affairs, community relations, customer service). But this traditional division is fading rather quickly as companies realize the commonality between, for example, the health concerns of a worker and the health concerns of a customer using the manufacturer's products.

Some large corporations have a large, centralized environmental function, with satellite operations spread across the globe at manufacturing sites. These satellites may have their own safety personnel and environmental engineers, but are unlikely to have, for instance, a toxicology staff. Ron Van Mynen, vice president for health, safety, and environmental affairs at Union Carbide Corp., notes that the cost of running medical-research laboratories runs into the millions of dollars each year; there is a compelling justification to centralizing all the analytical instruments in one place, rather than duplicating them in many places. Aside from such laboratories, however, there is ample room to decentralize environmental functions and to have specialists at various locations, or to have groups that handle the environmental affairs of specific divisions within a company.

This structure fits most readily with large corporations, whose hallmark is a large staff that maintains control over far-flung operations. The picture is quite different at small companies, where one individual may be responsible for worker safety, environmental management, community relations, and regulatory compliance. At even smaller (or less profitable) companies, one individual may be responsible for all these things, plus running the plant. Unquestionably, by raising the stakes with tighter environmental regulations, the federal government has made the plight of small manufacturing businesses much more strained.

Perhaps the most significant change in manufacturing management doesn't show on organizational charts. This is the new standard to which operational managers are being held to keep their plants running profitably. Waste-disposal costs and pollution-control spending are being charged against specific operating units—even specific plants. A sloppily run plant soon shows red flags to the corporate bookkeepers, who can see that environmental costs have been running out of line at one plant, relative to other plants in the organization. Thus, plant managers who normally would be most concerned with maximizing production now have a divided focus: Yes, they must keep production on track, but they must also keep environmental expenses under control. The temptation to cut corners by underreporting spills or by illegally dumping wastes is ever present. However, the legal consequences of doing so have become much stiffer, with some plant superintendents already heading for jail for environmental crimes.

"Green" Marketers

Biodegradable garbage bags. Laundry detergents made without enzymes. Cheeseburgers wrapped in paper, rather than plastic "clamshells." Welcome to the world of green marketing. Green marketing refers to the sale of products—mainly consumer goods—that have some implicit environmental benefit to them, with the expectation that this becomes a selling point.

"Environmental affairs have long been a priority in heavy industry, where producers of chemicals, oil, paper and steel have relied mainly on technical experts to bring heavy pollution under control at production plants" wrote J. Lublin in *The Wall Street Journal* (March 5, 1991, p. B-1). "Now, attention is shifting to the consumer sector, where environmental-policy executives often bring a background in product development, marketing or law."

For a long time, it has generally been true that business-to-business selling is a rather cut-and-dried affair, based on a product's cost and performance characteristics; presumed environmental benefits exist only if applicable regulations mandate them. In the consumer-marketing world, however, selling is based on perception as much as it is on reality. People will buy one cola over another on the basis of what rock-and-roll star is peddling the brand on TV. It is in this netherworld of perceived or claimed benefits that green marketing exists. Here are some examples:

■ Procter & Gamble, the giant household-goods company, is lambasted for producing disposable diapers, which use up more materials than reusable cloth diapers. P&G responds by pointing out that its diaper produces less water pollution, and besides, it can be composted into soil for growing trees.

■ Atlantic-Richfield Co. (ARCO), a major petroleum refiner, steals a march on other refiners by offering a high-octane gasoline with reduced "volatiles" (i.e., gasoline constituents that can evaporate and contribute to smog). While charging a premium price, they are able to expand market share.

■ Aveda Corporation, a Minneapolis firm, offers cosmetics that are, according to its advertisements, based on "the use of pure, natural, plant-derived compounds, in place of synthetic, petroleum-derived, or animal-tested products (some products may contain lanolin or beeswax obtained in a compassionate, careful manner)."

As these examples indicate, the range of pitches that marketers are making vary considerably. What is often missing from the equation, however, is a clear understanding of how and whether the existence and sale of a product will benefit the planet. "The real answer to the crisis of ecological degradation is not consuming appropriately, it is consuming less—a pitch you will not see coming out of Madison Avenue," wrote Andre Carothers, editor of *Greenpeace* magazine, in a guest column in *E, the Environmental Magazine* (March/April 1990, p. 73).

To be sure, there is an enormous potential for doing good for the environment on the basis of what consumer products one purchases. The challenge to marketers is to become environmentalists themselves, rather than to slap a green label on the conventional products they sell.

One example is Environmental Outfitters, a new company going into business in Santa Monica, California, in mid-1991. The prime mover of this business, which will sell construction products, is Paul Bierman-Lytle, an architect who has been at the forefront of designing and constructing buildings that minimize environmental harm through the use of recycled materials, energy-efficient equipment, and minimal impact on the land surrounding the sites. Environmental Outfitters will offer products whose synthetic-chemical content is low (the better to avoid the "sick building syndrome" that occurs when construction materials emit these chemicals). The store will be paired with a consumer-goods supplier that emphasizes green products, and the stores will be franchised and installed in shopping centers built according to Bierman-Lytle's design and construction principles. "I can't claim that I know everything there is to be said about the environmental effects of the products we sell, other than to say that we use them, and have not found reason not to," says Bierman-Lytle. "We call these products state-of-the-art, because that's what they are—they represent the current state of knowledge about building products."

From the outside, however, it is apparent that Bierman-Lytle is blazing a trail into two very tradition-bound industries: architecture and construction. By combining his firm's design services with a retail outlet for the products it uses and with a construction-contracting business he runs, he can provide a meaningful alternative to the typical project that is designed, specified, and built by three different parties with little coordination among them as to the environmental quality of the work being done. As much as any new line of green products, Bierman-Lytle's innovation represents a new marketing channel for customers of building products and services.

Traditionally, product marketing involves business training with experience in how products are advertised and sold. When the environmental effects of a product are added to the marketing mix, conventional marketers will simply seek the consulting services of a number of firms that have grown up to meet this need. The most accessible of such programs are Green Cross Certification Co. (Oakland, California) and Green Seal Inc. (Palo Alto, California). Both these firms seek to develop a "seal of approval" that consumer-product marketers can affix to their packages, much the way the Good Housekeeping seal of approval has worked its way into the marketplace. By going to a third party for this seal, product marketers hope that a verifiable environmental benefit can be turned into a profitable product.

The process is fraught with dangers on both sides, according to comments from both product marketers and environmentalists. These programs, as do private studies that diaper makers, plastic-packaging manufacturers, and others have already sponsored, depend on a "life-cycle analysis" of the product. Life-cycle analysts purport to review every environmental effect, from the time the raw materials of a product are first obtained, to its manufacture and distribution, to its form of reuse or disposal. That sounds simple in theory, but in practice it can involve a confusing mix of claims and counterclaims. When McDonald's switched from polystyrene clamshells to paper to package its hamburgers in late 1990, it cited studies showing that the plastic was a nonbiodegradable material, consuming crude oil. Then a Canadian scientist countered with a life-cycle analysis showing that polystyrene cups (and, by implication, other food packaging) were actually better for the environment (mostly because of the water pollution caused by paper making and the harm to forests that tree cutting entails). But the study did not note that less paper is used per unit of product; this one fact shifts the environmental impact the other way. In a larger sense, one can picture the same mess that has occurred in the health or toxicity claims made against foods on the basis of partial studies. The green-marketing movement could self-destruct in confusion and counterclaims. It is for this reason that many state government bodies, as well as the Environmental Protection Agency and the Federal Trade Commission have not attempted to establish this sort of program. State governments have been more forward, but mostly in the sense of preventing deceptive advertising.

In terms of employment, green-products manufacturing and marketing require business managers adept at translating the technical features of how products are made and used into environmental benefits understandable to the consumer. The issue of life-cycle analysis, which has implications for environmentalism far beyond green-products marketing, is a fruitful field of study for economists and other scientists. Already, a number of consulting firms have sprouted up and offer to perform these studies.

Recyclers

If people are spending money to throw junk into the ground and cover it with dirt, and you can find a way to sell some of that junk, how much money could you make? How high is up? That siren song has caused more than a few business entrepreneurs to crash on the rocks of failure as the complexities of making a go of recycling become clear. For in fact, even when junk is available, it can be expensive to extract it from the waste stream in a sufficiently pure form. And, while recycled products are often sold for a premium in today's environmentally aware market, it hasn't always been so—and isn't likely to be so for the indefinite future. Many customers of recyclers, believing that the product is of inferior quality and that it didn't cost anything to acquire anyway, insist on a discount relative to new materials.

Recyclers also bear the burden of close examination from environmental regulators. The reason for this is the mixed history of recycling in the United States. There is a giant loophole in such legislation as RCRA: A waste material, even a hazardous one, can be "delisted" with certain restrictions when it is clear that the material is a byproduct meant for recovery. Some businesses, taking advantage of this loophole, announce an intention to recycle a certain material—say, worn-out tires or automotive batteries—when, practically speaking, the main goal is to earn revenue by accepting the materials for disposal. When the junk-storage capacity at a site runs out, sometimes the "recycler" goes out of business as well, leaving it to Superfund and other governmental agencies to clean up the mess. All too often, EPA has found, a recycler is simply a waste hauler who, because the environmental regulations are not as strict for recycling as for disposing, puts a patina of "recyclability" over the dump. To meet these regulatory requirements, some recyclers are driven to the unbusinesslike practice of keeping no inventory of unprocessed material at their plants. What gets processed on one day is very likely to have been delivered the day before.

Nevertheless, recycling represents one of the most promising methods of reducing environmental harm, because the recycled material is not only removed from the waste-disposal cycle; it also displaces the demand for new materials. Before today's higher awareness of the environment, most of the recycling industry consisted of scrap dealers, who gathered paper, metals, rags, or chemicals such as solvents for reprocessing. These scrap dealers were concerned both with finding a sufficient source of supply and finding customers to pay for these products. By and large, they did not get involved with actual production of finished goods from wastes.

To this group was added, over the past decade or so, a new generation of technically trained people who sought to bring new technology to bear on recycling problems. Wellman, Inc., for example, is a Richmond, Virginia, recycler of plastics, which has carved a substantial market out for itself for being one of the first to base its raw material supplies totally on recycled material. Its sources include plastics manufacturers and, as the "postconsumer" (that is, recycling from trash) market developed, it has drawn on that source as well. The plastics are reprocessed and converted to polyester fibers, carpet backings, and similar materials; Wellman is now on the verge of becoming a billion-dollar company.

The latest entrant into the recycling markets are the materials manufacturers themselves. Industries ranging from paper to tire makers are feeling the pressure from their customers and state governments to protect their markets by recycling as much as other companies. Tin can makers, for example, have responded to the challenge of aluminum manufacturers by upgrading their recycling facilities. Plastics manufacturers have invested substantially in facilities for recovering bottles and other packaging. In the automotive industry, the use of plastics for body panels and other parts in place of steel is being maintained by corresponding efforts by the plastics suppliers to assure that their materials are as amenable as steel to recycling.

Postconsumer recycling is an extremely challenging task for government and industry. In many cities, garbage pickup is carried out by city employees, which puts the government itself squarely in the camp of potential recyclers. Cities are being spurred forward by the realization that space for landfilling the garbage is becoming restricted and is costing more than in the past. Scrap dealers, on the other hand, resent the intrusion of government into "their" market and have maintained, with some justification, that they offer a less expensive option to cities. The greatest challenge, however, is finding ways to encourage the recycling of materials discarded locally into products that can also be used locally. Newsprint, the paper used in newspapers, for example, makes a good potential application of this sort. Rather than having virgin paper hauled from across the country or across national borders to add to the stream that must be disposed of at higher and higher costs, city managers are seeking to encourage newspaper printers to recycle the locally obtained paper. The central question in all this is simply, who pays? If city sanitation crews perform the recycling, the city must invest in the process. If scrap dealers are reluctant to go after markets such as paper and plastics, city managers can seek to mandate the use of such materials. If the original materials manufacturers make no effort to recycle, governments may attempt to compel them to do so.

Green-collar engineers and production managers are needed to design and build the facilities that recycle postconsumer goods; new technology is desperately needed to develop ways to recycle those materials that have been heretofore forgotten. Recycling represents a profound challenge to local governments and, therefore, the study of public policy making is very pertinent. Finally, in cases where the recycling is accomplished by having consumers bring their own materials into a "materials recovery facility" (MRF, pronounced "murf"), strong, well-publicized efforts must be made, which create a demand for communications specialists.

Environmental Markets, 1990–91, by Richard K. Miller, projects that recycling will be a $60 billion industry in 1994. The field is one that readily lends itself to entrepreneurs, some of whom start from quite humble circumstances. Urban Ore, Inc. (Berkeley, California), for example, started with a research proposal in 1980 to excavate and recycle materials in an existing landfill in that city. When the funding didn't come through, as the founders recount in the *Whole Earth Ecolog*, Urban Ore simply reverted to scavenging, first at the landfill and then, when the landfill closed, at a transfer station that the city uses to collect garbage prior to shipping it to a more distant landfill. "Ten years later, Urban Ore generates over $600,000 per year selling reusable goods [and] employs fifteen people" (p. 91). The company rescues building materials such as discarded window frames or other wood, breaks up cement or concrete into gravel, and composts yard wastes.

Another entrepreneur, Gary Petersen, founded a company called Ecolo-Haul in 1972, when he was 24 years old. The company started as a conveyor of recyclable materials from municipal garbage in Los Angeles, with Petersen driving around town in a VW bus. It grew to include "reverse-vending" stores, where a customer could drive up, deliver recyclable materials, and receive cash on the spot. The company obtained Wall Street financing in 1982, and in 1988, sold out to Waste Management, Inc., for a cool $2 million (J. Goldstein, *In Business*, May/June 1990, p. 24).

The recycling industries growing across the country today represent one of the liveliest sectors of the environmental movement. The chance to start up or grow a recycling business, making a living while making a positive impact on the environment, is compelling. The field is fraught with some of the same dangers that exist among green marketers: companies with "Green" or "Eco-" or "Enviro-" emblazoned on their letterhead, yet which may be doing very little of actual benefit to the environment.

Disposers

After a lump of stuff has been made, used, reused, and rejected for recycling, it must be disposed of. The "ultimate" in materials processing is to return it to the ground, and Americans are prodigious waste generators; according to UNEP data, we produce the equivalent of 3.7 tons of personal and industrial waste per person, per year. It has been predicted, more than once, that future generations will garner much of their resources by picking through the garbage heaps we are creating today.

Three things happen with this waste: Much industrial waste is simply put back where it came from, that is, mining rubble is put back in the pit mine. Most garbage also goes into the ground, at a landfill, which is also where many industrial wastes end up. The final category is hazardous wastes, most of which come from industry and which now cost about $10 billion per year to get rid of. Many hazardous wastes used to be disposed of with municipal wastes, the theory being that they belonged together and would somehow neutralize each other. (This is another example of the old belief that "the solution to pollution is dilution"; in this case, municipal garbage serves to dilute the toxicity of the hazardous waste. Farfetched? In England, until quite recently, this was the standard practice.)

A variety of alternative treatments or pretreatments of wastes, such as incineration, composting, and biodegradation, is being developed. But inevitably, the wastes return to the ground. William Rathje, the "professor of garbology" at the University of Arizona, makes the case that civilizations can be tracked by the garbage they produce: In early, resource-scarce years, wastes are minimal and are continually reused. In the rich middle years, wastes accumulate and cities begin to move around or away from their trash heaps. In late years, as the social institutions decay, people essentially live in the midst of their garbage; it piles up until it practically buries the cities that made the culture thrive. Has the U.S. sidestepped this tradition in recent years, with the passage of RCRA, Superfund, and the other waste laws? The outlook is uncertain at present.

The central role in ultimate disposal, both for municipal and industrial garbage, is played by the city or state environmental officer. With municipal waste, city officials must decide whether the municipality should invest in a garbage incinerator, which burns the waste, simultaneously producing energy (which can be converted to electricity) and reducing the volume of the waste by about 80 percent. (The remaining ash, however, must still be landfilled, and this has become almost as much of a problem for the incinerator operators as nuclear wastes are to the nuclear industry.) Garbage incinerators are expensive; the capital cost, defrayed by charging a "tipping fee" to garbage haulers that bring their loads to the facility, works out to over $100 per ton in many localities. By comparison, tipping fees at landfills are $10 to $30 per ton. Why would anyone then build an incinerator? The reason is simply one of capacity—if there is no place to dump the garbage or location to ship it to, then incineration becomes the next best option. A famous garbage scow, the Mobray, was dragged from Long Island, New York, through the Caribbean, around the Gulf of Mexico, and back up the East Coast, finally coming back to New York where its load of garbage was incinerated in 1988. Other cities, especially along the Eastern Seaboard, are shipping their garbage 500 miles or more by rail and truck, to Ohio, West Virginia, and western Pennsylvania.

According to *The 1991 Resource Recovery Yearbook*, compiled by

Government Advisory Associates, a New York consulting firm, there are now about 120 municipal incinerators in the United States, and the total should rise to over 200 by the end of the century. Municipal incineration has been one of the fastest-growing of environmental businesses in the past decade. New construction, however, will begin to slow down dramatically in the next five years due to the doubts many people have expressed over the economic and environmental sense of incineration. Besides being expensive, these incinerators emit air pollution, including the bugaboo of industrial activity, dioxin. (A great number of studies have been conducted that cast some doubt over the supposed toxicity and carcinogenicity of dioxin, which is present in parts-per-trillion in incinerator exhausts, after those emissions have been scrubbed and cleansed by the air-pollution-control systems that incinerator operators are now obliged to install. But the fear of dioxin makes many people unmovable on the topic.) The NIMBY syndrome rises dramatically over the entire municipal incineration scene.

About a dozen firms, some of them quite large, are now in the business of building and operating municipal incinerators. While the incinerator itself is very similar to any number of manufacturing facilities, it's not up to a team of engineers and builders solely to get one running. Because a community or state is the customer, a complex interrelationship involving public and private financing, agreements with local utilities to purchase the energy produced, and territorial rights to garbage must be worked out. Thus, urban planners, economists, lawyers, financiers, and communications specialists get into the act as well. When the municipality decides to go the route of floating a government bond issue to pay for the construction, the entirety of American political machinery joins the fray.

New landfills are still being built, but with a much different character than those of the past. Federal and state regulations for these "sanitary" landfills require a system of heavy plastic liners, earthen cover, and an underground piping system to collect the leachate that inevitably forms. This business belongs primarily to civil engineering construction firms, who perform the site surveys, dig the foundation, and install the pollution-control equipment. While new landfill construction is somewhat less controversial than municipal incineration, the NIMBY syndrome shows up here as well.

On the hazardous-waste side, the situation is dramatically clear. People do not want a hazardous waste landfill or incinerator near them; nevertheless, new capacity must be found. The number of such landfills or incinerators is rather small—less than 500. Many industrial firms, lacking a service company willing to accept their wastes and wary of passing along a liability that may come back to haunt them, have installed their own facilities on site at their manufacturing plants. (Remember Rathje's point about living in one's own garbage?)

As with municipal incineration, the brutal conditions for making one's way in the hazardous-waste business have led to its domination by a handful of firms. These firms employ engineers (civil, mechanical, and chemical), laboratory specialists, and communications specialists, along with a predictable number of managers, lawyers, and financial controllers. Another element of the hazardous-waste business is its need for insurance: One of the requirements of RCRA is that a waste dump operator post a 30-year bond to cover the costs of maintaining and monitoring a landfill after it has reached capacity and been closed.

Special mention should be made here of a business that will continue into the future, but not indefinitely. This is the Superfund business—the process of winning a government contract (either from the EPA or from states) to characterize, and then clean up, an abandoned waste dump. Superfund was started in 1980 and only slowly has come up to speed in the number of contracts being let and the waste sites being cleaned. For the business community (and, most especially, for lawyers and insurance companies), much of the "action" of Superfund happens before a spade of earth is turned, in the investigations, lawsuits, and protracted negotiations over who will pay for the cleanup. EPA is authorized, in emergency conditions, to allow the cleanup to be carried out first and then to bill the potentially responsible parties (PRPs—see the definition on page 48). But more often, EPA seeks to reach a settlement with these parties first. The cleanup of an individual site can run into the tens of millions of dollars, so the stakes are very high.

At any rate, once a site has been selected for investigation, a paper history of it is tracked down in government documents, business licenses, and the like. Geoscientists and civil engineers are brought in to carry out a remedial investigation (RI) to see what wastes are present and what the hazards are and to determine the size of the site. (It is not uncommon for these sites to be decades old, to have minimal or nonexistent records, and to contain materials radically different from what the few records there are indicate). One aspect of this research is very much like the investigations a lawyer would conduct in preparation for a trial; the other aspect is highly technical, involving the digging of wells and extraction of geological samples. Along the way, a risk assessment is performed: Based on the presence and concentration of hazardous materials, public-health specialists evaluate the dangers present at the site and make a determination as to what risks are entailed in, for example, digging up the site. The culmination of this preparatory process is the feasibility study (FS), during which all the information that has been collected on the site, its history, and its risks are matched against a variety of treatment techniques. These can vary widely; a sampling is as follows:

- Bioremediation—using microbes, either already at the site or brought to it, to degrade the organic chemicals present; sometimes works, often takes years to complete.
- Capping and covering—bringing tons of clay and soil to cover over the site and reducing the amount of rainwater that percolates through it; relatively inexpensive because it leaves the contamination in place.
- Incineration—literally, burning the dirt to cleanse it; very expensive.
- Soil washing—takes out water-soluble materials, leaves some residue and polluted water.
- Solidification/fixation—a variety of methods that, essentially, turn the site into a solid block of cemented dirt.
- Slurry walls—a method of surrounding the site with a cement wall by injecting grout into the soil.
- Stripping—pumping up groundwater, exposing it to air to let volatile chemicals evaporate, and returning it to the ground.

The common element to all these techniques is their desperation. It boggles the mind to consider the intricacies of shoveling up dirt, trucking it to an incinerator, burning it, and then bringing it back to the site and dropping it back in the hole. Nevertheless, the sins of past environmental practices have come back to haunt today's business managers and government regulators, and something on the order of $10 billion per year is being spent, either by government or by private parties, in this onerous task.

Superfund has been a dramatic boon to two professions: geoscience and geological engineering. More is being learned about the chemistry of soil and how water and chemicals move through it than perhaps ever in history. For these sciences (geology and geological engineering, geophysics, hydrology, soil science, and oceanography), the timing couldn't have been better. Following the collapse of the mining industry in the United States, and the falloff in oil prices (both occurring during 1980–83), these professions had some of the worst employment records of any technical profession. "Earth sciences [were] the only category with a negative growth rate in a recent survey of graduate student enrollments by the National Science Foundation," noted *The Scientist* (February 18, 1991, p. 3). "The environmental field has caught on, and students are aware that there are a lot of job opportunities in environmental science," is how one educator summed up the situation in the article.

Superfund is not a business with an immortal future because the number of sites where past practices have caused environmental disasters is finite. However, only a few dozen sites had been completely restored by the end of the 1980s (even after 10 years of haggling and study). Some 1,200 sites remain on EPA's National Priority List, and hundreds, if not thousands, more sites remain uninvestigated. Superfund will be around for years to come.

Overall, waste disposal represents one of the most challenging tasks society faces, and one that will attract intense study, and money, during this decade. Public policy making, public administration, engineering, science, insurance, and public health are among the professions that are involved and will have a boost in employment.

Government

Federal Agencies

EPA was founded in 1970 in part to consolidate the environmental activities that were growing in the Department of the Interior, the Department of Health, Education, and Welfare (today's Department of Health and Human Services), the Department of Agriculture, and others. One might think, therefore, that EPA is the only employer. Wrong. If anything, it appears that EPA has grown to nearly 16,000 employees with no noticeable change at the other agencies. True, territorial imperatives have created areas where EPA is the leading, if not only, agency involved in environmental protection. But research, policy making, administration, and enforcement go on at all the other agencies, plus the Defense Department, the Department of Justice, the Council on Environmental Quality, and others.

In Congress, dozens of committees and subcommittees have a say in environmental policy, and when a major law such as the Clean Air Act, involving new taxes, financial regulations, and trade policies, comes before the two houses, practically every member of Congress becomes involved. This is as it should be, because it calls attention to the importance of the environment in the affairs of the nation.

For the job-seeking professional, government employment, whether at the national, state, or local level, has one major and one minor advantage and several important drawbacks.

The minor advantage is that these jobs can involve the individual in the most pressing issues of the day, helping to write history by getting laws passed, or by influencing the course of business and legal practices. Government is the work of the common good.

The major advantage of a government job is that, even while it is a relatively secure form of employment, it is also a powerful steppingstone to well-paying, vital jobs in the private sector. A lawyer with enforcement experience at a government agency has a perfect background for helping private-sector clients work their way through a thicket of regulations. Much has been written in recent years about "revolving-door agencies" in which this year's powerful regulator becomes next year's powerful lobbyist or dealmaker, and EPA and other environmental agencies have had this problem as much as the government at large. However, if you step away from the situation for a sense of perspective, you realize that this revolving door can be as much a force for good as for evil. Certain agencies, such as the U.S. Patent Office and the Internal Revenue Service, welcome former employees when they come before their offices for a case, because the expertise of these employees helps clarify murky issues and helps the agencies establish the proper procedures and ground rules. In this same way, a former government employee can help an inventor, perhaps, to get funding for an innovative new cleanup technology or help a wilderness preserve get the proper permits.

The drawbacks to government service are well known: relatively low pay, bureaucratic infighting, lack of control over policy. Government employees soon learn to develop a thick skin against criticism from the outside and to become adept at intra-agency and intragovernmental politicking. The frustrations, however, can be heart-stopping.

"It's well-known that our society usually makes decisions under duress, by engaging in litigation, which often serves no good purpose, especially when the environment is at stake," says Jonathan Lash, a former state environmental agency head (in Vermont), former nonprofit counsel (at the Natural Resources Defense Council), and now a law school head at the Vermont Law School (South Royalton), where he has established an environmental law program. "Other societies work out disagreements by negotiation or by having a mediator with the final word. It helps get decisions made faster." No one, however, contemplates a dramatic change in the United States' litigiousness.

Government service usually also means a constant lack of resources to do one's job. One former state environmental executive puts it this way: "It's a pretty heady experience to be a few years out of school, and to have primary responsibility over million-dollar projects, or over decisions that will affect thousands of workers. But then you notice how little help you are getting, and how difficult it is to get a decision made in your area of responsibility."

Many changes have occurred over the years regarding environmental policy and how it is carried out at all levels of government; more changes are in store for the future. There may come a day when industry is not battling regulators tooth and nail and when industrialists do not feel that regulators are biting their necks, sucking the lifeblood out of their companies, and when nonprofit organizations do not feel that government has sold out to industry with each new decision. The appointment of William Reilly as head of EPA was generally regarded as a maturing of the political process—a former environmentalist, appointed by the generally pro-business Republican administration (but there is already grumbling about a government sellout among environmentalist groups). In the meantime, government employment will continue to represent a form of work that can do great things for the environment.

Being the bureaucracy that it is, the federal government has fairly rigid hiring policies and pay grades. To get certain jobs, you must have the right academic degree. Pay is listed in a "General Schedule," which divides all employment categories into 18 GS levels. Each GS level, in turn, has ten pay "steps," which vary by about 30 percent from the bottom to the top. Typically, entry-level professional job categories begin at GS-5 through GS-11. In 1990, the base and top pay rates for each of these categories was as follows:

Job Grade	Lowest Pay	Highest (Step 10) Pay
GS-05	$16,305	$21,201
GS-06	18,174	23,628
GS-07	20,195	26,252
GS-08	22,367	29,081
GS-09	24,705	32,121
GS-10	27,206	35,369
GS-11	29,891	38,855

As the data in the next chapter will show, these pay rates are some 10 to 20 percent below what is normally available for environmental professionals in private industry. As a result, at various times, the federal government declares certain jobs to be in "shortage" category, which allows for higher pay. Engineers typically qualify for this higher pay, which adds some $5,000 to $6,000 to the schedule.

The Environmental Protection Agency. EPA acts as the primary enforcement agency regarding environmental standards. Other functions are to monitor environmental quality, to set new standards, and to conduct research. The agency divides the country into ten regions, with "strong" (EPA's word) regional offices in Boston, New York, Philadelphia, Atlanta, Chicago, Dallas, Kansas City (Kansas), Denver, San Francisco, and Seattle. In addition, there are four major research laboratories, in Las Vegas, Research Triangle Park (North Carolina), Cincinnati, and Corvallis (Oregon).

EPA is organized into "programs" (divisions), based on the type of environmental problem being addressed. These programs are research and development, air and radiation, water, solid waste and emergency response, pesticides and toxic substances, enforcement and compliance, and management. However, the multimedia nature of pollution, as it is coming to be understood, undercuts this organization. So far, EPA has not tried to combine different branches, but in R&D spending, for example, multimedia pollution research is the largest area in terms of fiscal-year 1991 funding.

Following the Reagan years of disdain for EPA and the election of an "environmental" president, EPA is beginning to reassess its place in the world. A Science Advisory Board of outside scientists and technologists recommended that EPA put new priorities on its programs and ranked global change as the number-one issue affecting human health. Other high-ranking topics were ozone-layer depletion, species extinction, and biodiversity. As yet, there is no "Rare Species Program" at EPA, but this may change shortly. Low-ranking categories include hazardous waste and oil spills—which happen to get very substantial attention in EPA's recent budgets.

Administrator William Reilly has expressed an interest in pushing EPA more in the direction of a policy-making body, rather than an enforcement one, but this kind of change will be years in the making. As that vision of the future comes to pass, EPA may find itself with fewer lawyers and investigators and more scientists and public administrators.

EPA hires approximately 800 to 1,000 new employees annually, mostly in the entry-level professional grades (GS-5, 7, and 9). Slightly over half have an engineering degree or other type of scientific training. According to Maureen Delaney, chief of the National Recruitment Program for EPA, "environmental engineer" is the most common technical occupational title. Following close behind is "environmental protection specialist," which includes graduates of chemistry, community planning, economics, environmental studies, geography, political science, public administration, or sociology. These specialists advise state officials, review or help prepare environmental impact statements, and administer the contracts that EPA lets.

Ms. Delaney goes on to say that hiring is decentralized at EPA; if you want a job in a particular region or at a certain laboratory, the best course of action is to apply at that location. The routine procedures for applying for a federal job are reviewed in Chapter 6; a list of addresses of EPA offices is in Appendix C.

Department of Agriculture. As of 1988, there were a total of 121,000 employees in this executive department. The majority of these employees work in agencies or service bureaus that help farmers raise product quality and quantity. Agencies within Agriculture that employ green-collar workers include:

- The Forest Service, which manages almost 200 million acres of National Forest land. (The National Forest system resides in Agriculture because Theodore Roosevelt put it there in 1905, taking it from the Department of the Interior.) The Forest Service employs a broad range of biological scientists and foresters and keeps the National Parks system functioning for visitors. Student employment is also available.

- The Soil Conservation Service, which began in the Dust Bowl days of the 1930s. Soil conservation is inextricably intertwined with water-resource management. Engineers, soil scientists, and soil conservationists are among the occupational titles.

- The Agricultural Research Service is the R&D arm of Agriculture, employing over 8,000. One-third are scientists and engineers, mostly with Ph.D. degrees. Occupational titles include chemists, food/chemical/agricultural engineers, biologists, geneticists, and agronomists.

Department of Commerce. Commerce is a far-flung entity, with divisions ranging from the International Trade Administration to the National Oceanic and Atmospheric Administration. Agencies with green-collar aspects include:

- The National Institute of Standards and Technology, formerly the National Bureau of Standards. NIST is the weigher and measurer of last resort for the country; its original purpose was to maintain the integrity of product-measuring methods. Today, it has branched out into a wide variety of manufacturing and scientific research. Engineers and scientists from the bachelor to Ph.D. levels are hired into such occupational

titles as chemical engineer, electrical/electronics engineer, fire prevention engineer, mathematician, and physicist. There are 3,000 employees in Gaithersburg, Maryland, and Boulder, Colorado.

- The National Oceanic and Atmospheric Administration (NOAA),
  which is currently basking in the afterglow of the research that went
  into the discovery of the ozone hole above the Antarctic. NOAA is
  also the manager of the National Weather Service—meteorologist to
  the nation. Bachelor's and master's degree holders in the full range of
  geosciences are employed, as well as biologists, oceanographers, and
  mathematicians. With the upcoming debate over global warming,
  NOAA figures to be centrally involved in resolving the knotty technical issues.

Department of Defense, Army Corps of Engineers. These "military"
engineers are carried on the books of the Pentagon and stand ready to go
into action on an emergency basis during wartime or natural disasters and
spend the rest of the time dredging or restoring rivers, lakes, and the coast-
lines. Civil and mechanical engineers, along with parks administrators and
natural-resource managers, round out the environmental aspects of Corps
work.

**Department of Health and Human Services, Public Health Serv-
ice.** There are a variety of agencies within this service that bear directly on
environmental health and risk assessment. The Public Health Service itself
is an operational division, putting thousands of public-health workers in offices throughout the country. Health specialists, physicians, biologists, and
computer scientists are among the specified job backgrounds. Within PHS,
the Agency for Toxic Substances and Disease Registry was founded (in
1985) to compile data on environmental hazards related to Superfund and
RCRA regulations. At the Centers for Disease Control, a Center for Environmental Health oversees state-run health maintenance and prevention
programs.

National Institutes of Health. More than 13,000 researchers and health-
care professionals are employed in this agency to develop new technology
for curing illness. One of the institutes that bears special emphasis here is the
National Institute of Environmental Health Sciences (located in Research
Triangle Park, North Carolina), where work is done on identifying the environmental factors that affect health. Nurses, physicians, engineers, medical
technologists, and statisticians are among the occupational titles listed.

Department of the Interior. Interior has 120,000 employees, and many of them are involved in green-collar work. At the same time, many Interior activities have been pointedly criticized by the rest of the environmental community. All too often, when the issue is a dispute between private-industry exploitation of natural resources and wilderness preservationists, Interior has almost automatically sided with private industry. Much of the organizational philosophy of Interior is oriented toward use of natural resources, rather than preservation of them; Interior is usually one of the more prominent plums for political assignment when a new administration comes into office. Among the most significant agencies there are:

- The Bureau of Mines, which conducts long-term research on mining technology, ore processing, mineral use, and mine-site restoration. Metallurgist, chemist, geologist, economist, and mining engineer are among the occupational titles listed.

- The Bureau of Reclamation, which was responsible for the construction of most of the major dams that exist in the American West. More recently, BuRec has been developing an expertise in creating habitats for migratory birds and fish at the reservoirs its dams have created. Civil engineer, electrical engineer, geologist, and hydrologist are among the titles listed.

- The U.S. Geological Survey, which started as the organization that made maps of the unexplored American West and has become the primary agency for mapping surface features, evaluating groundwater resources, and performing studies to mitigate the effects of earthquakes, floods, and volcanoes. Civil engineer, cartographer, hydrologist, and geophysicist are among the occupational titles listed.

- The National Parks Service, the front line in maintaining the 77 million acres of national parkland that America has reserved. Many historical sites are also under the purview of Parks. Archeology, business administration, forestry, and physical science are among the college majors Parks seeks for its professional staff.

- The U.S. Fish and Wildlife Service conserves wildlife and regulates its hunting. Managing the 31 million "recreation visitor-days" (i.e., one person spending 12 hours at a National Park for hunting or fishing) that Americans devote to hunting and fishing is the responsibility of this agency. A variety of biological specialities are the desired educational background of job candidates for the Service.

■ The Bureau of Land Management is probably the most controversial of Interior agencies. Its stated goal is to manage lands under "sustained use" and "multiple yield" for competing interests such as mining, forestry, and recreation. It has 12 state offices, 58 district offices, and 143 resource area offices, mostly in the American West where it manages 270 million acres of public lands. A long list of academic specialties is sought for the Bureau, including mining, wildlife biology, civil engineering, petroleum engineering, cartography, and land-law examination. It runs a student intern program and provides cooperative education.

Department of Labor, Occupational Safety and Health Administration. OSHA came into being at about the same time as EPA, and while the latter's problems seem to grow from year to year, OSHA has quieted down considerably from its earlier days. OSHA can also point to generally downward trends in workplace accidents as proof of its effectiveness. Industrial hygienist is the main occupational title.

National Aeronautics and Space Administration. NASA *is* space, and with the winding down of the Shuttle program, the use of it to transport earth-gazing satellites that will monitor resources and environmental quality is now one of the administration's stated goals. It is a technology-driven agency, requiring large staffs of aeronautical, electronic, and mechanical engineers, as well as computer scientists and mathematicians.

National Science Foundation. NSF doesn't "do" science in its own laboratories, but it does sponsor research across all ranges of science and technology. Thus, scientists are not recruited, by and large, on the basis of their scientific prowess; the accounting and business management skills necessary for contract administration are most desired.

Nuclear Regulatory Commission. NRC is in the unhappy position of watching the industry for which it was created stagnate—commercial nuclear power. Still, the rise of concern over global warming, combined with the research and technical development needed to dispose of nuclear wastes (including those from the federal nuclear weapons complex, over which NRC has no jurisdictional say, but for which it might provide substantial technical expertise) will keep NRC hard at work in coming years. Specified areas of academic training for employment at the commission include most types of engineering (including environmental), health physics, and metallurgy.

State Agencies

By various laws, each state must have some designated office or function that addresses the environmental issues the federal government has a hand in. Some states have very aggressive environmental programs, with large staffs that do their own research or initiate new legislation. Other states have a bare-bones office, which is there to respond to a federal query, but which does little else. States have had different historical approaches to environmental activities, too. In the older states of the East, the growth of large cities early in the nineteenth century led to departments of public health and sanitation. In the Midwest, the dominance of agriculture lent that flavor to state-level environmental activities. In the West, some states have made environmental activity an appendage of preexisting natural-resources-development concern.

The 1960s counterculture charge to "think globally, act locally" has led to many states taking on a role in national or even international environmental affairs, by voting to ban a specific consumer product or to regulate environmental hazards a certain way. According to the constitutional separation of powers, states are free to write more restrictive rules than the federal government does, but not to write more lenient ones. State legislatures today consider hundreds of bills annually on subjects ranging from the banning of plastic hamburger cases to mandating the conversion of corn into fuel for automobiles.

"State-level work is probably the most satisfying for the environmentally minded," observes one veteran of the scene. "At the federal level, too much is out of your control, or set by another's agenda. At the local level, it is very difficult to get around local power centers. But at the state level, you can institute a new program, and see it get off the ground through your own efforts." He adds, though, that the state government executives will always be forced to compete for limited resources from other branches of the government; there is never enough to go around.

The state of California is far-and-away the leader in environmental activity at the state level. It is so far ahead, in fact, that it frequently passes legislation or sets policies that the federal government later adopts. Environmental activities are divided among a number of agencies: The Air Resources Board, the Energy Commission, the Department of Health

Services, the Office of Statewide Health Planning and Development, the Water Resources Board, and the Water Resources Control Board.

Within one of these boards, there are a variety of divisions. In the Air Resources Board, for example, there are divisions of compliance, research, stationary source (of air pollution), monitoring and laboratory, and mobile source. Hazardous wastes—a big concern in California—are regulated by a Toxic Substances Control Division within the Department of Health Services. Overall, the state employs hundreds of civil, environmental, chemical, and geological engineers, and hundreds more public-health specialists, doctors, chemists, public administrators, lawyers, and economists. In order to be hired, the Personnel Office uses a test and an interview. Good grades, volunteer work, and technical expertise count heavily.

Local Government

Local government is, if you will, the business end of the political process. City and county officials are the ones who actually buy and operate waste water treatment plants, get sanitary landfills sited, get the garbage collected, and send out the payment-due notices for public services. Jurisdictions vary from city to city and state to state as to what types of jobs and environmental responsibilities belong to the locality and which to the state. There are certain instances, such as pollution-control along a river, where state-level (or even multistate) responsibility is necessary, because the river is used by every locality along its length. Large cities often have their own staffs of engineers, public-health officials, and environmental administrators, especially for air pollution, which tends to be worse in cities than in the country.

Perhaps the most dramatic contrast is not between state and local government but between urban and rural government. A number of studies have shown that rural communities tend to have the poorest governmental resources and depend on state-level government to meet most of their needs. Rural areas, too, suffer disproportionately from the presence of waste dumps, both of municipal and industrial wastes, and the concomitant degradation in groundwater quality.

Don't expect to find research facilities allocated at the city level, except of the most rudimentary kind. Do expect to find jobs that involve significant responsibilities in interacting with the public, whether it is as a parks administrator or as a city official sitting in at town board meetings where water rates are set.

Business Services

Science, Engineering, and Architecture Consulting

The unifying theme to these business services is that they are "technical." The results these consultants provide are written in terms of chemical concentrations, soil samples, or construction specs. The sections following this one deal with the nontechnical business services, such as advertising or insurance.

Without question, the most profitable sector of the environmental business today is the provision of these technical business services. Revenues for hazardous waste services, for example, reached some $6 billion in 1989, and were growing at a 20 percent per year clip. To put that in context, the U.S. economy as a whole grows by 2 to 4 percent per year, and most businesses feel good about doubling that rate—that is, growing by 4 to 6 percent. Only the superhot computer and biotechnology industries, in recent years, have been able to achieve growth rates in excess of 10 percent per year for several years running. And the technical consulting business has grown at twice that rate.

Employee turnover is a serious problem in many companies in this field, not because the money is bad (it's not) or because the managers are tyrants (they'd soon be out of business). It's simply a supply-and-demand question: There are so many more jobs than there are qualified people to fill them. Once a company gets desperate enough, it will pay through the nose for the right job candidate.

"I get a call two or three times a month from a headhunter offering me an attractive new assignment," says Ravi Krishnaiah, an environmental engineer at Brown and Caldwell Consultants, Inc. (Pleasant Hill, California). And it's not because the young engineer is an old pro at Brown and Caldwell that people are calling; he's only been there a year. Previously, he worked as a sanitary engineer in the Department of the Ecology for the State of Washington and had extensive experience with Superfund sites. The combination of government service and some private-industry experience now makes him an extremely tempting target for headhunters.

Another indicator of the health of the business can be found at The Environmental Resources Management Group, Inc. (Exton, Pennsylvania), which was founded by Paul Woodruff in his home in 1977. In the first year of operation, it earned $318,000—then $5 million in 1982, $22 million in 1985, and $65 million in 1988. Employment has gone from 1 at its founding, to 60 in 1982, and hit 500 in 1987. Such growth rates are unsustainable, and if a company is planning for growth by hiring lots of people, and the growth doesn't materialize, the company's finances could quickly be wrecked. But such uncertainties are part of the territory in these business services.

Just what services are being offered? Among the engineering firms, the work can be broken down almost according to the appropriate federal law:

- Superfund investigations and cleanups.
- Environmental audits (to assure compliance with RCRA) at chemical companies and other hazardous waste producers.
- Geoscience services, to investigate groundwater conditions, to study a site prior to commencing construction, or to examine a site prior to its acquisition.
- Waste minimization, to provide expertise in reducing the production of wastes.
- Risk assessments, to evaluate the potential liabilities of a cleanup action or other environmental activity.
- Regulatory compliance, to help a company meet statutory requirements for recordkeeping or permit applications.
- Underground storage tank management, to examine these tanks (found throughout industry) for leaks and potential damage and to bring them into compliance with new regulations.
- Litigation support, to assist a law office in making a case or defending a client.
- Policy analysis for government agencies.

The list could go on and on. It is apparent that these services call on a quite varied pool of talents. The story of The Environmental Resources Management Group, starting with a man in his home to a 500-person office in little over a decade is unusual in its string of successes, but is not unusual in its goal to gain breadth of many different aspects of the engineering-consulting business. Most of these companies, if they want to expand at all, want to expand horizontally, developing or acquiring expertise in a variety of engineering services. In this way, the enterprise as a whole is insulated from the ups and downs of economic, business, and regulatory trends. When EPA's Construction Grants program was throttled back in the early 1980s, for example, it caused many engineering companies that were dependent on water pollution control to suffer.

Many of these companies come out of the tradition of civil engineering construction, through which public works such as highways and city buildings have been constructed. But the growth of environmental business has led to mutations of the business, with some firms now offering high-tech solutions to industrial problems, consulting on regulatory compliance, or providing emergency-response services (which, in turn, require the expertise of fire, health, and safety personnel). Banks and insurance companies now routinely call on their services to evaluate the environmental quality of a plot of land or a factory complex.

Engineering is naturally the dominant profession among the staffs of these firms. Specific engineering disciplines include civil, sanitary, environmental, chemical, mechanical, geological, and industrial. In addition, other professionals include such scientists as geologists (and most other earth scientists), biologists, chemists, foresters (conservation scientists), health physicists, and medical doctors. Lawyers, accountants, communicators, marketing managers, and salespeople round out the list of professionals that are being employed.

The engineering-consulting firms, accustomed to being "purchased" for their expertise, readily buy the services of other technologists, most notably, analytical laboratories. Consider for a moment the vast amount of data that must be obtained to understand what is going on at a Superfund site:

- Composition and concentrations of dozens of chemicals.
- Quality of water and measurements of its flow.
- Soil composition (i.e., clay, humus, stones) and permeability and strength of underlying strata.
- Weather patterns, such as wind directions and precipitation.
- Effects of countermeasures, such as the use of microbial action to reduce toxicity.
- Biopsies of workers at the site or local residents.

Again, the list could go on and on. These studies are simply the first step: to characterize the conditions at the site. More analyses are performed as remedial work progresses and then as a permanent monitoring system is installed. It is no surprise, therefore, that analytical laboratories earn an estimated $1 billion per year, strictly for environmental work.

These laboratories perform tests that are not too different from what goes on in a hospital or university laboratory. Chemists and biologists take samples, prepare them, run the sample through an instrument, and record the results. However, the sheer volume of testing that is done demands that a laboratory be successful at automating as much of the testing process as possible—letting machines run multiple tests, for example, or developing systems for automatic data logging and analysis. Environmental work has helped change the several hundred laboratories that perform this work around the nation, encouraging the development of computer programs and sophisticated electronics to carry out the tests.

The much greater degree of sensitivity that environmental research requires also leads to innovations in the types of instruments. As will be discussed in Chapter 7, James Lovelock, the British scientist, was able to notice for the first time that CFCs were able to reach the ozone layer of the earth only after a new instrument had been invented; it was this realization that spurred other scientists to examine the implications of CFC buildup in the atmosphere.

Another group of scientific consultants is those who are developing new methods of cleaning up past pollution or preventing the generation of new pollution. Much of this work is the conventional new-product development that goes on at the major equipment suppliers to industry—those that manufacture purification systems, chemical separators, and biological reactors. But the dramatically new types of problems that the environmental business has encountered are helping put a green collar on university researchers, individual inventors, and innovative manufacturing firms that can exploit their own environmental problems when they develop a solution that can then be marketed to other firms.

A recently published study, "Advanced Separation Technologies," from Frost & Sullivan, Inc., a market-research firm in New York, gives some examples of what is occurring:

- Bio-Recovery Systems, a Las Cruces, New Mexico firm, takes academic research on the use of microbes to extract precious metals from ore and applies it to a pressing problem: the removal of heavy metals such as cadmium or chromium from waste water. The live microbes are embedded on a matrix of gelatin beads, and the contaminated water is trickled over a column of the beads. The result is water clean enough to be reused in such applications as electroplating or electronics manufacturing, plus the elimination of a hazardous waste and a significant health hazard.

- Scientists at Battelle Memorial Institute, a research organization in Columbus, Ohio, discover that the application of a combined electric/

acoustic-energy field to wet sludge accelerates its drying in a filter. The process is accelerated, allowing the same filtering system to process larger amounts of sludge in a cost-effective manner. A commercialization joint venture is then organized with an equipment manufacturer.

■ Glitsch, Inc., a Dallas, Texas, manufacturer of equipment for the chemical industry, purchases the licenses to a process developed in the United Kingdom that combines two very conventional materials-processing techniques in a very unconventional manner. The two processes are distillation (essentially, a still boils off liquids to separate them) and centrifugation (a process whereby a centrifuge is used in biological laboratories to concentrate cellular matter at the bottom of a test tube by spinning that tube extremely rapidly about a central axis). "Centrifugal distillation" offers the potential to pack a tremendous amount of distillation capacity into a small unit, a pill-shaped, motor-driven cylinder that drastically speeds up the separation of compounds that normally takes place during distillation. Such a machine could be carted to a waste dump, or installed in a crowded factory building, to remove organic contaminants from waste water or even to distill very refractory materials.

"It is a truism of process engineering that 90% of the work is to get the last 10% out of a stream," concludes the Frost & Sullivan study. "Never has this been truer than in the 1980's, as a new round of environmental legislation has mandated that [parts-per-million] levels of contaminants must be reduced to the [parts-per-billion] range." The point here isn't that scientists and other technologists are inventing new things—that process goes on all the time. The difference is that, with such a vast potential market of applications, the environmental science community is developing a new set of tools that will be able to be tested in a rigorous fashion. Innovations that succeed in the environmental arena have the potential to "cross over" as new manufacturing technology that will change the way things are made today.

Architecture is often linked to engineering through the nexus of construction: What an architect designs, a civil engineer builds. In architecture, too, a realization is dawning that to be environmentally conscious is also to be business minded. Robert Berkibile, FAIA, a Kansas City, Missouri, architect, is playing a major role in driving home this realization. As chairman of a steering group for the Committee on the Environment of the American Institute of Architects (AIA, of Washington, D.C.), Berkibile is pushing for the development of design guidelines that would emphasize the environmental safety of the buildings, a harmonious use of the land surrounding a building site, and an awareness—backed by clear-cut evaluations—of the environmental effects of the building materials that are chosen for a project.

The first two points are not controversial, if you stop to think about them. Buildings can be environmentally unsafe; the "sick building syndrome," in which office workers are made ill by poor ventilation or by the outgassing of certain chemicals from new construction materials, is well known and is the subject of study at EPA and other organizations. (Ironically, EPA underwent a sick building syndrome episode a year ago at its Washington headquarters; new carpeting that had been installed gave off so many fumes that the building was temporarily evacuated.)

Creating harmonious designs is also noncontroversial; architects spend a good part of their training learning how to make use of the space surrounding a building. It may never come to pass that any AIA architect will ignore an environmentally unsound construction project; however, by directly influencing the value judgments architects make, Berkibile and his group members hope to ultimately influence society at large as to wholesome construction.

The third point is the one sure to generate the most controversy. With backing from EPA, AIA is embarking on a multiyear project to evaluate the environmental soundness of building materials. "I got started on this several years ago, after my firm designed a downtown skyscraper that would have richly paneled rooms for a law firm moving into the building. One of the considerations for appropriateness of wood building materials is whether they come from a sustainable-development site, as opposed to a one-of-a-kind site that would be destroyed if the wood is harvested," Berkibile says. "When we had an open house to celebrate these new law offices, the cabinetry subcontractor brought photos of the actual tree that was cut down, in the African savannah, to make the paneling. When we all saw that, it was a shock, and I vowed not to let that happen again."

AIA's goal is to develop a life-cycle analysis of building materials, ranking them according to the harm they cause when mined or harvested, then used, then discarded. A preference would be given to those materials that have the least harmful effect. Manufacturers of such materials, understandably, are somewhat upset; they would prefer that the piece of aluminum made from bauxite mined from the Amazonian rain forest, for example, should be treated the same as a recycled aluminum soft drink can. But Berkibile will press on. "One of the reasons that this project has EPA funding is because it reflects EPA's desire to get out of the enforcement business and into waste prevention and policymaking," he says. "Another reason is that architects have grown too comfortable, believing that they could proceed with designs regardless of the environmental impact of their choices. We have a lot of education before us to do."

Legal Services, Insurance, and Risk Management

Environmental Law. What's this? A market study that "shows key practice areas in nation's largest law firms" finds that "environmental work led the pack, with a more than three-time increase in the number of firms now reporting this area among their top five revenue-producing practices"?

"It's true," avers Kathy Mitchell, a vice president at Sherry & Bellows, a marketing consulting firm for the legal profession. Periodically, the firm surveys the law firms for the Of Counsel 500, an industry tabulation. For the 1987–1990 period, environmental work jumped dramatically for the firms surveyed, so much so that some firms depend on it as their largest income generator. Some of the firms that Sherry & Bellows included in their poll showed a more than 100 percent increase in their environmental business.

While it is clear that environmental work has increased in recent years, it is rather shocking to see how fast it has grown in the legal professions. At the American Bar Association (Washington, D.C.), members have developed a variety of new interest groups within their organization. According to Courtney Leyendeker, a staff member of ABA's Standing Committee on Environmental Law, there are at least 40 different committees or groups within the ABA organization that have an interest in environmental law. "This aspect of the law touches nearly everything today, so it's no surprise that all these groups are up and running," she says. Some are decades old, but the number of them has increased dramatically in recent years.

Almost by osmosis, law students have caught on to the growing need for environmental training. At the Vermont Law Center (South Royalton), director Jonathan Lash says that the number of students applying has shown "explosive growth." The school offers legal training on two levels: a full-fledged law degree or a master's degree in environmental law that prepares the student to work in the environmental compliance departments of corporations or of the government. "When you consider that the new Clean Air Act runs 780 pages long, you can see that legal issues on the environment have become very complex," says Lash. "This master's degree provides graduates who are versed in the newest legislation."

The legal services that are required by major corporations are diverse. Simply to deal with a new law or regulation sometimes requires a negotiation with EPA or another environmental agency. When an infraction is recorded, the case may be disputed for years by the party accused by the regulatory agency. In other aspects, lawyers are needed simply to keep the wheels of commerce rolling. When an engineering company bids on a cleanup project, it may require lawyers to review the bid application, making sure that it represents a service that the company can indeed offer and that it meets the regulatory demands of the funding agency.

One of the more gripping aspects of environmental laws is that several of them have built-in provisions for settling a court case. A nonprofit group, for example, can take EPA or a deep-pocketed corporation to trial, and if the nonprofit wins, it can then collect from the funds that EPA has set aside for such legal settlements. Another element in this picture is the actuality of being charged with felonious "environmental crimes" and being obliged to go to jail if the law is enforced. At Pitney, Hardin, Kipp, and Szuch, a Florham Park, New Jersey, law firm, representing corporate officials has led to the tripling, followed by doubling, of the number of staff working these cases. Lawyers, of course, are not only being used on the side of corporate executives. Nonprofits, citizens' action groups, private adjudications, and other interested parties depend on lawyers to make their cases.

"Our environmental practice grew out of the general litigation department about ten years ago, due to the increased volume of work that was being done in this area" notes Clyde Szuch, managing partner at Pitney, Hardin. "What is interesting now is how environmental concerns cut across many traditional legal specialties—business clients must be advised on environmental audits; banks seek advice on valuations of industrial property, based on its environmental condition; even trust and estate law is affected, such as when contaminated land is part of an estate." Szuch says that none of the firm's clients has required defense in a criminal trial, but that the risks of this happening grow with the increased regulation. "We're environmentalists," he adds, "because when called on, we show a client what has to be done, and point out the risks of not doing it."

The final element of the environmental law picture is what goes on behind closed doors, when new environmental regulations are sought. Most of the lobbying organizations in Washington, for example, have legal staffs that go into action when a new law is being written. Some 800 environmental measures were put before Congress in 1989, and while few of them pass, many of them do take up hundreds of hours of lawyers' time.

Insurance and Risk Management. The insurance needs of businesses and individuals dealing with environmental issues have also undergone change. In particular, companies involved in cleanup work or waste disposal must be able to post bond that their work will be completed satisfactorily. The insurance industry has long experience with loss control and risk management, and both these fields are being stressed in environmental activities. "Risk management" is a term that denotes both a management responsibility and a specialized set of knowledge, according to Dr. Dan Anderson, chairman of the Actuarial Science, Risk Management and Insurance Department of the School of Business at the University of Wisconsin (Madison). "In corporations, the title of Risk Manager is coming to be the general term for the responsibility of plant safety and security, insurance coverage and accident prevention, and environmental affairs," he says. But in his school's programs, risk management is a study of how losses occur and how they can be prevented, in the context of empirical knowledge about industrial accidents. "A general liberal arts background is appropriate for this field," he says. The course work has a fair bit of mathematics, but does not involve heavy quantitative analysis the way actuarial science would.

Mark Johnson, vice president at an environmental firm and a former student of Dr. Anderson, testifies to the value a risk-management degree had when he entered the consulting field in the mid-1980s. "I used to work in healthcare for the disabled, but got frustrated with the slowness and uncertainty of getting results through my efforts," he recalls. "I came to Washington, not really expecting anything—certainly not environmental work. But with my background, very quickly I got a job as a policy analyst, and grew with the firm." That company, now called PRC/EMI, is a McLean, Virginia, EPA contractor and business consulting firm. Mr. Johnson manages a staff of 320.

Market Research

Any field in which changes are occurring rapidly and large amounts of money are at stake (either to be paid or to be earned) is quickly going to attract market research specialists who provide the data that business managers in the field require to make decisions. In the context of the green-collar work force, this market research is needed by the consulting and business-service firms looking to build their business volumes up and by the consumer-goods firms that want to gain an edge in the evolving "green" marketplace.

"The marketing skills of many of the environmental consulting companies are minimal to nonexistent," declares Gail Brice, president of Brice EnviroVentures (Newport Beach, California). Her firm, which employs her experience as a marketing and new business development manager for a number of industrial and consulting firms, exemplifies how the need is being filled.

"In many cases, new, fast-growing environmental firms are run by a bunch of engineers who have had little marketing experience," she says. A company attempting to market, for example, a new groundwater cleanup technique, cannot simply announce its availability and wait for the phone to start ringing. The first step could be giving speeches about the technology at industry conferences. The next could be the "courageous client"—a firm with the need for a technology such as what is being offered, but which can take the time to allow test work to be done, and whose environmental problem would not become worse if the technology didn't work. With a field demonstration like this under its belt, the company could then approach governmental regulators for evaluation of the system under guidelines that have already been established. "The goal is to be declared a 'best demonstrated technology,' because then you are officially considered a viable technique for any number of other cleanup projects," Brice says.

Such a "long-cycle sell" requires skillful marketing and market research: knowing when to seek government approval for a technique, rather than continuing its refinement, and knowing how to estimate the potential size of the market to get a sense of the technology's future potential.

In "green" consumer-goods marketing, market research looks very much like that of any other consumer goods, with the exception that a great emphasis is placed on life-cycle analysis. The "greenness" of a product is determined by how little pollution or environmental damage it causes, from when its raw materials are collected to when it is finally disposed of, after being used and possibly reused. Market researchers continue to probe American consumers to find out just how environmentally committed they are, because very often, a recycled or "green" product costs more than its conventional equivalent. Will the consumer be willing to pay a premium for this environmental improvement? Market research provides an answer.

In sum, the business services that environmental organizations depend on are very much like those that serve traditional businesses: consulting, legal and insurance, market research. Relative to other sectors of the green-collar work force, these consultants probably put in the longest hours and receive the highest pay. But those salaries exact their own cost. "Consulting work is very fast-paced, with crazy deadlines," says Jackie Spiszman, who worked in the field in the past and now is a project manager in the California Department of Health Services. "It's a field for risk-takers, and the people who are successful are those that enjoy the 'game.' It's very lucrative, but you give up a lot of your personal life."

Communications, Publishing, and Advertising

It would be hard to imagine a social activity or line of work that requires so much constant communicating as environmental issues. From the basic task of educating the public about what issues are important, what is known, and what risks are at hand, to getting the vote of approval before local zoning boards for an environmentally related business, the need for high-quality communications is universal. And whereas most types of public relations or information-transmitting functions in society are voluntary ones, in many cases these communications are mandated by law in the environmental field.

Two cases in point: When an environmental engineering firm is called in to clean up a dump site or to install pollution-control devices at a factory, it is now required to share with the public the minute details of how it will go about its work. Community "right-to-know" legislation, now common in most states, gives the local populace the opportunity to make its own evaluation of the risks and hazards of the intended project.

Similarly, manufacturers that make or use hazardous materials are obliged legally to share with their workers the details on the risks and proper handling procedures of the materials in question. While many companies have done this voluntarily in the past, as a far-sighted gesture to improving working conditions, today it is a matter of law.

The communicators who carry out this work generally do not have a technical background, the better to understand the information needs of the general public or the company employees. The work demands a willingness to learn continually and to convey written or oral information in an understandable manner. These types of communication occur in an arena where mountains of information are being transferred from one party to another on a continuous basis. Public relations specialists sing the praises of their clients before the press and the financial community. Journalists grapple with complex technical questions associated with new laws and industry's response to those laws. Advertising agencies, especially those with "green" clients, find the right words to win the interest of ever-more-concerned consumers.

All of this, of course, occurs in the charged atmosphere of environmental activism, political ideology, and profound fears over the health of the members of communities across the nation. "The best preparation for a public meeting over a hazardous waste project," says John Schlatter, only half-jokingly, "is to give the communicator a blindfold, light a cigarette, and ask if there are any last words before the execution." As community relations manager for Bechtel National, Inc. (an engineering and construction firm in Oak Ridge, Tennessee), Schlatter has participated in some extremely intense public hearings over environmental projects. Being able to convey the appropriate information, in a knowledgeable manner, goes a long way to defusing these situations, he says.

"I'm a translator," says Stephanie Reith, community relations manager at Donohue & Associates, Inc., a Chicago, Illinois, engineering firm. "I take highly technical information and put it into understandable language that the public can deal with." Having worked for nearly ten years in this field, Ms. Reith is now a practiced hand at this work, but she notes that she studied art criticism in college—a subject as far away from the work she does now as could be conceived. The "keystone" to her work, she says, is the ability to write and think clearly. "If you can't write well, you won't be a good communicator, regardless of the medium you work in."

Ms. Reith and a group of like-minded individuals have recently formed a professional organization for their field, called the National Association of Professional Environmental Communicators (see Appendix A for address). A critical issue that will be addressed in this organization is professional ethics. It would be easy to imagine an environmental communicator simply glossing over the details of a client's project, but experience shows that this type of PR often backfires when the truth becomes known. Honesty and trustworthiness, which must be built up over time with journalists and community leaders, are the goals.

One typical recipient of the information Ms. Reith provides is Craig Howson, editor of *Waste Tech News*, an industry newspaper. "I come from a daily-newspaper background, with a degree in journalism," he says. "I was a government reporter, but this is more fun—you get to deal with an enormous variety of technical, political and business issues." Howson's publication provides this information to a readership of managers and purchasers of environmental equipment and services, in an easily digested format, a "niche" that he says was unfilled before his publication was started about four years ago. "Being a journalist in this field is similar to how the various publications are organized," he says. There are publications solely for the technological specialties, as well as for environmental activism, political issues, and others. "You have to find your own niche."

Perhaps the most dynamic form of environmental communications occurs at nonprofit organizations—both those that represent radical environmentalism and those that speak for industry or other commercial-interest groups. Like a politician running for office, these organizations seek to present a compelling message before the public to sway opinion one way or another on environmental issues. A good example of this occurred in California during 1990 over the fight for "Big Green," a voter referendum to dramatically increase the regulation of pesticides and to prevent the exploitation of natural resources. Coalitions of environmental-activism groups lined up against industry and Chamber of Commerce interests, as voters were inundated with a flood of reports, position papers, and broadcast advertisements. Big Green lost at the polling booths, but California remains at the forefront of environmental legislation, and polluting companies feel an increased pressure to comply with the activists' agenda of demands.

The main communications medium where objectivity is usually not assumed is advertising. Indeed, a growing number of companies have been hauled into court by aggressive state attorneys general when advertising claims were shown to be false or misleading. However, this question of objectivity does not negate the need for meaningful, effective advertising of environmentally benign products and services. As shown earlier in this chapter, more and more businesses are learning how to adapt to customer demands for new, less polluting products. Companies whose products are losing market share due to the better environmental quality of what their competitors sell go green in a hurry. And the acceptance of these better products may be the dominant means of improving the environment.

Typically, advertising agencies are organized around three professionals: the account executive, the creative director, and the copywriter. The first coordinates marketing information and strategy with the client; the latter two are responsible for creating the message that will successfully sell a product or service. While many more dollars are spent on television advertising than on print media, the team must be prepared to deal with both.

Travel, Tourism, and Recreation

Each of these industries is only secondarily a "green" one, although both the economic and business impacts of tourism and recreation are substantial. In fact, many more radical environmentalists would argue that too many facilities for tourism and recreation in the wild eventually cause it to suffer.

On the other hand, it is definitely true that tourism and recreation are among the primary "recruiting grounds" for conservationists and others who are environmentally minded. After having gone on a camping trip and seen life in the wild, who does not think differently about wilderness and its need for "development" or "management" (words that have all too often been euphemisms for "exploitation")? In fact, one of the growth areas in travel in the early 1990s is what has come to be called "ecotourism"—the arrangement of package tours to rain forests or wildlife reservations, where the backpacking or sunbathing is intermixed with lectures on indigenous ecosystems. Nor is the environmentally pristine the only destination: Unexpectedly popular tours include the Boston Harbor (a sewage-choked bay that figured so largely during the 1988 presidential election) and the site of the Chernobyl nuclear disaster in the Soviet Union.

At any rate, the nation's network of national, state, and private parks, and the amount of time people spend in them, are growing steadily. According to federal statistics, the number of visits to all federal recreation areas grew from 6.4 million visitor-hours (their unit of measurement) in 1980 to 7.5 million in 1988 (*1990 Statistical Abstract*, p. 222). Expenditures at national parks, specifically, rose from around $700 million in 1980 to almost $950 million. As more and more efforts like that of the Nature Conservancy and The Trust for Public Land, which seek to preserve the wild condition of land throughout the country, take effect, more land will become privately owned, but publicly used, parks.

One of the key occupations among parks employees and wildlife professionals is that of "interpreter," a job that is part teaching, part tour guide, and part conservationist. For people to enjoy a wilderness or recreational experience, guidance is usually important, and the interpreter is called upon to provide it. The need for interpretation also demonstrates the intertwining of *teaching* environmental concepts (see "Teaching" below) at the same time that one is *practicing* environmental work. Interpreters work both at private and public parks and conservation areas. Competition for jobs is usually intense, and pay is not very high.

Similar popularity is enjoyed in hunting and fishing—a $55 billion activity in the mid-1980s (*1990 Statistical Abstract*, p. 234). As was pointed out earlier, hunters and fishers are somewhat at odds with environmentalists who seek uncompromising preservation of wilderness. Nevertheless, they have been at the forefront of supporting the cleanup of rivers and lakes and the prevention of land overdevelopment.

Wilderness and species preservation is one of the growth areas in this field, and it is growing internationally as well as domestically. Private organizations such as the Nature Conservancy or The Trust for Public Land have developed sophisticated campaigns to elicit contributions from landowners, and they help local groups develop the necessary expertise to manage the resource. This expertise includes forestry, water-resource management, ecology, and biology. Planning, fund raising, tax law, and estate planning are among the important business skills these efforts require. In the international arena, "debt for land" swaps have become a popular means of preserving wilderness areas in Third World countries, whose foreign-debt burdens are huge. By pledging to preserve a wilderness, a portion of that country's debt is forgiven. These highly complex, politicized endeavors require the highest levels of expertise in international banking, diplomacy, and resource planning.

An organization like The Trust for Public Land, in particular, also practices land restoration. Land that has served one purpose—say, a railroad right-of-way through a scenic part of the countryside—is converted to "wilderness" by removing the traces of human construction and by managing the re-introduction of appropriate plants and wildlife. Somewhat the same purpose is served by a variety of private and public organizations that manage wildlife for hunting. Here, the skills of fish and wildlife management, agronomy, ecology, and biology are important.

Teaching

No one can say with any degree of precision how many "environmental" students there are, but everyone agrees that their number is increasing dramatically. This is occurring even during a time when the number of students applying to colleges and graduate schools is unchanged or declining, due to the smaller demographic group now of college age in the United States. (Overall enrollments may be maintained, or may even rise, depending on whether more teenagers decide to apply to college.)

There are about 250 undergraduate colleges and universities that offer programs in environmental studies and about 50 schools that offer some form of environmental engineering (which is sometimes subsumed into civil engineering, or given another name like sanitary engineering or environmental science). Each passing month seems to bring news of a new program being started up, often due to intense student demand for one.

An even more dramatic surge is occurring at the professional (post-college) level. New programs are being started by schools to provide training certificates and/or master's degrees for new workers in the environmental field, or to enable professionals with one type of training (for example, engineering) to learn about another field—say, risk management. Some of these programs are of significance only to the people taking them, but others have by now developed enough of a track record that employers know to look to them to gain new staff.

Edward Demos, an adjunct professor at the University of Denver, started such a program in the fall of 1990 to offer a master's degree in environmental management. After consulting with local industry leaders (who would be potential employers), developing a curriculum, and winning support for teaching facilities at the university, Dr. Demos announced the program in early 1990, expecting 30 students to get the program off to a solid start. He got 150. "We're walking on air right now, because we see a pent-up demand for this type of schooling, and a need among employers for graduates with the training we can provide."

Similar stories can be told in other parts of the country. One common theme to many of these programs is their interdisciplinary nature. Environmental management depends on people who know both biology and physical science, engineering and business, health and accident prevention, and so on. This often makes a program difficult to set up when schools of science, engineering, health care, or business are not in close connection with each other. But this trouble is by no means insurmountable.

The issue is further complicated by the differing philosophies of how environmental concepts should be taught, especially to pre-college students. Writing in *Environmental Communicator*, the professional journal of the North American Association for Environmental Education (May/June 1990, p. 9), Rick Mrazek, an environmental educator in Lethbridge, Alberta, notes that the conventional model, which he calls a "traditional inductive-deductive or theory-oriented approach," is but one way. Others are "historical-hermeneutic," which revolves around exploratory case studies; "critical-theoretic" which calls for an eclectic mixing together of scientific study, political, ethical, and economic theories, and other appropriate bodies of knowledge. In all these models, the chance to learn from practitioners ("interpreters") in actual work settings can be an important element in teaching.

For the near term, there is a definite need for green-collar teachers. For the longer term, it appears that environmental studies will move up in prominence among the schools of science, health, and public policy.

Nonprofit Organizations

This category contains a range of groups that would often be uncomfortable being in the same room. These include the environmental "public interest" groups, industry-sponsored trade associations, and privately or publicly funded "think tanks" that develop policy stances with a variety of political or economic points of view. The reason these groups are covered together has nothing to do with their politics or policies, but with the nature of the work done at them.

What is that work? Essentially, all of these groups gather information, analyze it, formulate a policy, and then publicize their findings. They also serve as information sources to the media or the public at large (which also serves as a form of publicizing). Some conduct or sponsor scientific research. Some of them are authorized to be lobbyists in government. Some of them have sizeable staffs of lawyers and attempt to influence public policy by hauling either government agencies or industries into court. Many of them, especially the public-interest ones, do extensive fund raising. But the core activities, again, are information gathering, analysis, policy making, and publicity.

Without question, the number-one skill needed to succeed at this work is communications ability. Even a brilliant scientist performing innovative research must go beyond the laboratory to represent his or her work to the public at large; a skillful lawyer must be able to hold sway in the court of public opinion, as much as in the courts of law.

Many people in nonprofits enjoy the tremendous rush of causing society to change direction or revealing an unknown fact that changes the public's perception of what is going on in the world. The Environmental Defense Fund, for example, was instrumental a number of years ago in convincing Pacific Gas & Electric, a big California utility, to institute a program of conservation, rather than simply building yet another power plant, a move that foreshadowed similar programs around the country. Sometimes, beleaguered officials in government secretly praise the efforts of the nonprofits that are socking them with lawsuits in the public courts, because that might be the only way to make progress on a thorny environmental issue. And while trade groups all too often toe the line set by their industrial sponsors, very often they bring technical expertise and experience to the table that is lacking among other participants in the debate. Trade groups often work for the public good, but in a gray area where antitrust law could become a factor. The Chemical Manufacturers Association, for example, is a large, well-funded lobbying organization for the chemical industry, but it also administers the Chemtox program, whereby a hazardous-materials team is on call around the clock to respond to accidental spills and highway or rail accidents.

Most of the best-known environmental organizations actually have quite small permanent staffs, although the organizations' total memberships can number in the hundreds of thousands. Table 2 shows the summary details about some of the larger groups.

The efforts of these organizations are greatly magnified by their volunteer supporters and contributing members. The Sierra Club, for example, has more than 500,000 members.

Table 2 Profile of Leading Environmental-activist Organizations

Organization	Permanent Staff	Annual Budget (million $)	Year of Founding
Conservation Foundation	65	3.6	1948
Cousteau Society	100	11	1973
The Environmental Defense Fund	75	8	1967
Environmental Law Institute	50	4	1969
Greenpeace USA	150	50	1979
National Audubon Society	25	32	1905
National Wildlife Federation	550	70	1936
Natural Resources Defense Council	125	11.7	1970
Sierra Club	275	28	1892
Wilderness Society	112	11	1935
World Wildlife Fund	100	14	1961

Nonprofit work is hard and sometimes heartbreaking. Many of the organizations that are funded by private donations lead a hardscrabble existence, with highly trained professionals earning sub-par salaries. But the public limelight and the opportunity to do good can make up for these shortcomings.

CHAPTER FIVE

Environmental Professions

Following is a (edited) job advertisement that appeared in *The New York Times*, January 13, 1991:

> **CORPORATE MANAGER, ENVIRONMENTAL AFFAIRS**
> Leading international electronics manufacturer is seeking a dynamic, strategic person to fill the newly created position of Corporate Manager, Environmental Affairs. . . . Reporting to the Corporate Executive level . . . this leadership position will have primary responsibility for the development and implementation of Corporate Environmental Policies; Management of Environmental Programs and Projects; and Coordination of Plant-level Environment-related activities.
> The ideal candidate will have at least two years corporate experience in a manufacturing company, with four to six years experience directly related to environmental issues. . . . This includes in-depth knowledge of federal environmental regulations including RCRA, CAA, CWA, CERCLA, SARA III, and TSCA. . . . Bachelor of Science in Chemical Engineering or related area preferred. . . . Competitive salary and excellent flexible benefits including matched savings, profit sharing, pension plan, and dental plan. . . .

This (edited) job advertisement ran in the *Environmental Opportunities* newsletter, January 18, 1991:

> Small company offering natural history safaris, rafting & fishing to adults/families; activities occur on the Kenai Peninsula & in Denali National Park. Positions available for skillful, high energy, & people-oriented outdoor leaders mid-May through mid-Sept.: natural history trip leaders & guides, rafting guides, fishing guides, trip drivers,

cooks, housekeepers, . . . Salaries range from $800–$1,600/month (plus gratuities). . . . Tent housing w/nominal deduction for meals. Call/write for application packet.

What kind of green-collar worker do you want to be? What kinds of skills do you have or want to acquire? What kind of life-style do you want to lead, and what salary do you need to support that?

The preceding chapters gave a feel for what organizations or employ- ment opportunities exist in today's green-collar work force. Throughout, certain assumptions were made about the academic training, experience, and skills of the workers involved. This chapter will lay out the details of what academic degrees or work experience are desirable. The emphasis will be on entry-level positions available to the new or recent college graduate. Where available, information on certification, necessary skills, and day-to-day work responsibilities will be detailed.

In today's dynamic job market, it is a tossup whether an individual con- siders himself or herself allied more closely with an industry (or type of com- pany) or a profession. That is to say, is an environmental-impact writer at heart a writer (and could just as easily be a journalist, communicator, or community-relations specialist), or an environmental policy analyst? And could that writer transfer his or her skills to siting a waste-disposal facility, set- ting up a recycling collective, or regulating industrial pollution?

There's no right or wrong answer to these questions; it depends very much on the individual. In terms of the U.S. workplace today, though, the answer falls very much on the side of alliance to a profession rather than to an industry. During the 1980s many large corporations cut their staffs re- lentlessly, seeking to reduce overhead to an absolute minimum. At the same time, the traditional loyalty felt between employees and their employers shriveled; the individual worker, left to his or her own devices, shifted em- ployers by means of demonstrating professional skills that were applicable to different types of industries.

For this reason, some of the information in this chapter pertains to the profession as a whole, and not exclusively to the green-collar aspects of that job. Specifically, where applicable, data will be cited from the U.S. Bureau of Labor Statistics' (BLS) *Occupational Projections and Training Data*. Ac- cording to the 1990 edition of this publication, the U.S. work force as a whole will increase by 15.3 percent, from 118 million to 136 million. A pro- jected increase higher than this is generally a good sign, indicating that de- mand may be overreaching the supply of new graduates or those with appropriate experience. However, it is also true that in most professions (green collar or not), more jobs are opened up by retirements, transfers, and the like than by demand changes.

Keep in mind the interdisciplinary nature of environmental work. Large companies have highly trained individuals in many of the job titles listed here. Other companies or other employment sectors (government, nonprofits) seek an individual who has command of several specialties. Keep in mind, too, the evolving nature of the green-collar world: The high-demand jobs of today will be replaced by others in coming years. It is a wise career strategy to choose a type of work that has transferable skills. The specialist gets paid a lot of money, but when the job is over, he or she often disappears.

Spotlight on the Environmental Manager

There is one job title that deserves special attention because it is brand new and it lies at the very heart of today's green-collar work force. This is a job that usually goes by the title of Environmental Manager. Most of them work in business or industry, but a sizeable fraction work for consulting organizations, waste disposers, and recyclers. The environmental manager is at the operating end of things—making sure that pollution-control technologies are working correctly, that community relations are being maintained, that workplace safety is at its highest, and that products being brought in for recycling meet whatever quality specifications are demanded for both the process and reused product.

This sounds like it pertains mostly to manufacturers, such as the chemicals industry or metals refiners. That is probably true today, but that is only because the need is greatest there. Eventually, all types of commercial enterprises, and governmental organizations, will employ these professionals. Consider this: What type of enterprise does not produce any solid, liquid, or gaseous waste (that is, what organization has no problem with solid waste disposal, water pollution, or air pollution)? What organization does not have workplace safety standards to be maintained and insurance policies to be continually evaluated? And what organization, either through the products it offers to the marketplace or the materials it uses during the course of its business, does not face recycling opportunities? Far-sighted companies that one would not expect to have significant environmental worries are already establishing such positions, including airlines, telecommunications firms, and retailers. In the future, most companies will either be creating such a position or contracting with a consultant to provide it on a semiregular basis. (This applies primarily to larger companies; smaller companies will make environmental management one of several functions for a senior manager, and at very small companies, as is almost always true, the proprietor will wear all the hats.)

Training for environmental managers usually starts with a technical degree in engineering, engineering technology, or chemistry. In the future, better established environmental management programs at colleges will provide the training directly. But the engineering or science degree is only the start. Today's environmental managers add to their base of skills by studying environmental policy, public administration, or business administration. Health sciences and occupational safety are being incorporated as well. Depending on how energy-intensive a business is, the study of energy technology may also be a component of the training.

The job market is crying for people with such training, but they are very hard to find. As supply reaches equilibrium with demand, the high salaries will moderate. (Many are earning $50,000 or more with just a few years of work experience, while the environmental affairs managers at some Fortune 500 companies earn well in excess of $100,000.) Following are summaries of the more notable green-collar workers.

Green-Collar Professions

Agronomist

Agronomy is in a ferment of new technology and new methods these days. While the traditional duties of agronomy continue—developing methods of planting, raising, and harvesting foodstuffs more economically—a host of new concerns are being raised, both within and outside of the agricultural sector of the U.S. economy.

For example, environmentally conscientious farmers and farm scientists are looking more closely at "low-input" agriculture. This term refers to the abandonment, or at least drastic reduction, of fertilizers, pesticides, and intensive plowing. As the channels for marketing environmentally "pure" foodstuffs develop and strengthen, more farmers are looking at low-input agriculture as a viable alternative to conventional farming practices. (Anyone with a smattering of technical training is taken aback by the usual term that is given to these pure products—"100 percent organic"—as if crops grown with fertilizers and other chemicals are "nonorganic.")

Agronomy also is important in various types of nonfood growing, such as lawn care, landscaping, and land remediation. On the latter point, literature from the American Society of Agronomy states that "agronomists . . .

are conducting research on the potential problems associated with recycling solid wastes into soil [and] are exploring techniques for reclaiming and revegetating drastically disturbed lands such as toxic waste dumps" (*Exploring Careers in Agronomy*, 1989, p. 10).

For the profession as a whole, the outlook is clouded by the severe downturn in employment caused by the slumping agriculture markets in the United States. Although food production has not shrunk, employment in the field has. Enrollments in colleges have shown a corresponding decline. However, many agronomy educators predict drastic shortages of suitably trained professionals in coming years.

Air Quality Engineer

Developing technologies both to analyze and control air pollution requires relatively sophisticated engineering skills. The new Clean Air Act, passed in late 1990, requires a complex set of countermeasures to air emissions. These include the reduction of so-called "fugitive" emissions (which are the small leaks that occur in factories, rather than what comes out of a smokestack), restrictions on automotive exhausts, reduction of sulfur dioxide and nitrous oxides from utilities and other power plants, and more study and control of indoor air quality. Simply to confirm the existence of some of these types of pollution is not a trivial task.

Most engineers involved with power production have a mechanical engineering degree, but such studies as chemistry, chemical engineering, or environmental engineering are also appropriate. Jobs occur at engineering consulting firms, in state government and at EPA, and among manufacturers and utilities.

Biologist

The study of biology is one of the core preparations for environmental work and, since nearly 45,000 students graduate annually at all academic levels in this field, it is also one of the most popular college majors. The job title of "biologist" is rare because there are so many well-established specialties: aquatic biologist, biochemist, botanist, geneticist, zoologist, microbiologist, and others. (Ecology, generally a part of most biology programs, is treated separately here.)

Many biologists find work in environmental laboratories, working basically under the same conditions as chemists (see entry)—running tests and making evaluations. The key difference, of course, is that the biologist may be dealing with living creatures rather than test tubes and electronic analyzers. The "bioassay," by which the health effects of a contaminant in water or air are measured by its effect on fish, microbes, plants, or other life forms, is a common feature of environmental studies.

Biology is the gateway to a great variety of other professions mentioned here, including ecologist, wildlife manager, toxicologist, industrial hygienist, and biostatistician. The entire realm of health care beckons: physician, epidemiologist, health-care researcher. Biologists tend to have advanced schooling; roughly one out of eight go on to obtain an advanced degree in biology, and many others go on to medical school or other forms of advanced education.

Microbiologists are finding fruitful opportunities in bioremediation work, by which naturally occurring microbes, or genetically selected ones, are used to detoxify underground aquifers. (The first genetically engineered microbe to win a patent, in fact, was one under development for bioremediation.) Many companies that service the waste-water treatment industry also employ biologists.

Average starting salaries are in the range of $18,000–24,000, depending on the type of employment (lower for straight lab work, higher for design or research). Projected growth is 15 percent.

Biostatistician

Biostatistics (and the more or less synonymous term "biometrics") is the statistical analysis of health condition, usually of human health. Biostatistical work has become especially important in the risk assessments that are carried out in conjunction with pollution-control operations, or with the determination of the cause of disease or injury. The split between a knowledge of biology and a knowledge of statistics is fairly even, but in most schools, the biostatistics are learned in the biology department. An advanced degree is often desirable. The heavy quantitative aspect of biostatistics makes it imperative that computer skills be well developed.

The statistical interpretation of health effects is one of the most controversial in medical circles. The connection between chemicals and cancer, for instance, has been in a state of turmoil for years over the validity of the Ames test as a measure of risk. Most biostatisticians work for the federal government, although there are many opportunities in the health-care field in association with hospitals or contractors that provide this statistical service.

Chemical Engineer

A chemical engineering degree gives its holder broad entree into nearly all aspects of the environmental field except research (which would be open to the Ph.D. chemical engineer) and medical specialties. The irony is that chemical engineering enrollments have declined steadily since 1984, and half as many students are graduating now with a B.S.E. degree as did then. Salaries, not surprisingly, have shot upward and now start at around $35,000. In some states, a professional engineer (P.E.) license can be obtained, which is helpful for any engineer involved in public works projects (where the possession of a P.E. license is often mandatory).

The chemical industry and engineering/consulting firms are the highest bidders for chemical engineers, and with the supply shortage, are taking most of them. The heart of chemical engineering education involves an understanding of "unit operations"—the basic steps by which a material is transformed from its raw to finished state. This knowledge applies readily to projects like Superfund remediation (where one or a few unit operations are used). The familiarity with chemistry enables the chemical engineer to be adept at workplace safety, waste-handling or recycling activities, or air- and water-pollution control.

According to federal data, there are about 49,000 chemical engineers, and projected growth over this decade is 16.4 percent.

Chemist

The laboratory is the traditional home of chemists, but the need to gather data from polluted dumps or sick buildings is bringing them out of the lab. A steadily growing revolution in laboratory instrumentation enables chemists to perform many varied analytical tasks, such as sensing underground deposits, measuring the air quality high in the sky over a factory, or analyzing the combustion byproducts of automobile exhausts.

Laboratory employment, however, does not pay very well—about $24,000 annually—so many chemists go on for advanced degrees or seek employment opportunities outside the laboratory. Academic training in environmental studies would make the chemist a good candidate for working in consulting or assuming a managerial role in a laboratory, with somewhat higher pay.

Some chemists' positions are more involved with day-to-day operations, such as at a water-treatment works, where the varying quality of incoming waste water calls for variations in the chemical treatments provided.

And in industry generally, many chemists do research on improving production processes, reducing wastes, or inventing new products. Some of these jobs go by the title of "environmental chemist."

BLS says that there are about 80,000 chemists working currently; the single largest employer is the pharmaceutical industry. Overall growth is projected at 13.7 percent—average for all professions.

Civil/Environmental Engineer

The civil/environmental engineer is probably the most common type of engineer at work in the environmental arena today. Overall job prospects are in a slump currently, due to the overheated construction business of the 1980s. By some estimates, we will be well into the second decade of the twenty-first century before all the unfilled commercial space in downtown cities is rented. Environmental engineering is usually offered in conjunction with a civil engineering degree, as a concentration or option, especially at the graduate level.

So, it's a good thing for the profession that the environmental business is booming. Civil engineers are best known for moving earth, and the Superfund projects across the country provide a natural application for that skill. Depending on the program attended, there are possibilities for combining civil and environmental training; in this case, the engineer learns more about groundwater, solid-waste disposal, and similar subjects. A large proportion of civil engineers in private practice seek to obtain a professional engineer's license, which is mandatory in certain circumstances for public works design and construction.

Average pay for starting B.S.E. civil engineers is around $28,000; in government employment, it starts at around $27,000.

Community Relations Manager

Community right-to-know laws now require engineering firms or manufacturers to reveal detailed information on the expected risks of a cleanup action or for preparing for emergencies near chemical plants, utilities, and other factories. The conduit through which this information is passed is the community relations manager.

Essentially a function of the public relations department at large corporations, community relations management calls for a rather complete set of communication, diplomatic, and technical skills. Many lack an engineering or science degree, but in order to be successful, such managers must be able to cope with the flood of scientific data and analysis that are issuing from Washington and state capitals.

Community relations managers find employment at consulting firms, especially those involved with the cleanup work paid for by the federal Superfund, and at manufacturers, especially chemical companies and others with a heavy burden of pollution-control responsibility. The skills learned and applied in performing community relations work lend themselves readily to environmental activism, public relations, and other communications-dependent professions. There hasn't been, as yet, a thorough survey of job titles, salaries, and related issues for the profession, but if the patterns of public relations professionals set the trend, starting salaries should be in the vicinity of $22,000–24,000. At the top of the heap, corporate environmental affairs managers can command salaries of $60,000 or better. As the recently organized National Association of Professional Environmental Communicators (see Appendix A) gets under way, more such data should become available in the near future.

Computer Specialist/Database Manager

The computer revolution of the past 25 years has certainly influenced the environmental field. Besides being a general-purpose office tool that any executive should have a passing familiarity with, the computer is used in highly complex research simulations of environmental conditions, as a monitor for the vast array of manufacturing processes that must be tightly controlled to prevent accidental emissions, and as a recordkeeping device for the flood of paperwork that environmental projects create.

The multidisciplinary nature of environmental work means that no single profession or field of study has all or even most of the answers for an environmental project. Thus, there is a need for various specialists to pool their knowledge in a database to which all interested parties have access. EPA runs a series of such databases on existing regulations, preferred remediation technologies, hazardous chemicals, and many others. Every year, EPA makes available the "Toxic Release Inventory," a summing-up of all the pollution generated by industry; a computer-readable version of this inventory is used by local and national environmental groups to pinpoint what factories, in what cities, are a target for legal action.

On the research side, the subject of "modeling" is a dynamic focus for studying environmental issues. To predict, 50 or 100 years from now, the effects of global warming, computer scientists and physicists have developed a variety of "general climatological models" (GCMs) that simulate the actions and reactions of sunlight, weather, rainfall, and pollution on the Earth's climate. Today's regulatory decisions are being made on the basis of the output of these GCMs.

Database management—the gathering and sorting of data and information—is a key computer specialty throughout industry, consulting firms, and government. Graduates of computer science, information science, or library management can find employment opportunities among consulting firms, government agencies, and large corporations. The research applications of computers usually call for technical training in addition to computer science, in fields such as mathematics or physics.

The number of computer professionals throughout the U.S. work force is huge—some one million at least—according to federal data. The projected growth rate is high as well, pegged at 50 percent or better over this decade. Computer specialists who want to work in the environmental field should plan to take some courses in science, public affairs, or health administration to complement their computer courses.

Contract Administrator

Overseeing the work of hired contractors is a critical skill for most public agencies, which are given a budget and the responsibility to accomplish a set of goals by hiring private companies to carry out the work. The pattern is set by a construction company, which is hired to design and build a structure while meeting a time and financial budget. There is quite a bit of construction-related work performed at various governmental environmental agencies, but environmental work also entails many other types of contracts, such as scientific studies, policy studies, surveys of technical specialists or the general public, and more.

The contract administrator's duty is to clearly understand the guidelines under which a contract is being offered, convey those guidelines to the contractor, and then see to it that they are met. The skills involved include cost estimation, an understanding of regulatory codes, and a familiarity with how the work is usually performed. For these reasons, contract administration is seldom an entry-level job. More often, one gains experience by working for a contractor or as an assistant administrator.

Even so, one of the truisms of the workplace shows up dramatically in contract administration. In the private sector, it could be years before a newly hired engineer or business manager gets to administer contracts; at public agencies, that could be an employee's responsibility in just a couple of years.

The job is often specified such that an engineering background is required, but frequently, other types of training, including general business administration, are appropriate. Expertise in cost estimating is also valuable.

Earth Scientist

The earth sciences comprise geology, oceanography, soil science, atmospheric science and meteorology, and certain specialties in physics and astronomy. The earth-based specialties (geological engineering, paleontology, geochemistry, geophysics) suffered a decline during the 1980s, as enrollments and job opportunities plummeted due to the collapse of the oil-exploration business. Nevertheless, in 1988 40 percent of such scientists worked for the oil and gas industries; the second highest total was the federal government (17 percent).

Atmospheric science underwent a big jump in interest during the 1980s as a result of the chlorinated-fluorocarbon (CFC) debate; this new interest will probably continue, if not strengthen, as a result of the global warming debate. These concerns translate into Ph.D.-level jobs in government and academia. Teaching positions are expected to open up further as this decade wears on, with more college-level teachers reaching retirement age and a projected growth in college-age students.

Ph.D.-level geoscientists get a starting salary of around $50,000 in private industry. In federal government employ, the starting pay is around $35,000. At the B.S. level, starting salaries for the two categories are, respectively, $23,000 and $17,000.

Dramatic things are in store for earth scientists due to the increasing use of "remote sensing" technology, that is, data from satellites. This technology is already being used to locate buried hazardous wastes in Eastern Europe (the data is from U.S. satellites), and its use in mineral exploration, agriculture, and natural-resource management is growing.

Ecologist

People who call themselves ecologists are generally involved in research, usually in the academic arena. Workers who apply ecological principles to environmental needs are foresters, soil scientists, wildlife biologists, and others.

As a study of living things, ecology is a subset of biology (see "Biologist"). It differs in that it is the study of systems of living things—an ecosystem. This is one of the more critical technical needs in the environmental field these days, as nature groups, industrialists, and wildlife lovers seek to preserve wild habitats and slow the extinction of endangered species. Mid-1980s estimates of entry-level salaries by the Ecological Society of America were around $20,000, with Ph.D. holders commanding $55,000 or more. The study also showed that nearly two out of three researchers held academic positions. In academic research, the trend is usually toward specializing in the ecology of one system (or even one organism in a system), whereas in the working world of wildlife preservation, the need is usually for ecologists that can deal with varieties of systems.

Environmental Protection Specialist

This job title is favored by EPA to describe, primarily, an entry-level position whose occupant can oversee a Superfund contract or monitor a manufacturer's recordkeeping routines. A knowledge of applicable federal laws and regulations is essential. A scientific background is desirable but not essential, while such quantitative skills as accounting or statistics are a boost.

For EPA itself, the work essentially entails reviewing contracts and reports, making site visits to monitor actual conditions, and making judgments based on the health and safety criteria that have been established for various toxic compounds. The work is primarily administrative in nature. EPA rates it at GS 9–11, or roughly $25,000 to $30,000 per year.

In the private sector, this responsibility translates into an environmental compliance specialist, who is likely an assistant to an environmental manager. This specialist keeps tabs on new regulations and is responsible for writing mandated reports. In the consulting industry, a engineering or technical degree is strongly advised; in private industry, generally, a technical degree should suffice. Starting pay is around $28,000.

Forester

Here's a situation pulled from recent newspaper headlines: Medical researchers discover that an extract, called taxol, from the bark of a species of yew tree is a potentially powerful cancer drug. The older the bark, the better the yield of the drug. The yew trees are not extremely rare; on the other hand, they do not cover whole mountainsides.

What does the forester bring to an issue like this? First, how many such yew trees are there, and where? A survey can be made. Second, if the choice is to harvest them (recognizing that each tree harvested diminishes the stock), how should it be done? Third, if there is a reasonable chance of domesticating the tree and growing whole fields of it, how is that to be done? The forester has many of the technical tools needed to answer these questions.

The yew tree is only the latest example of how humanity has found a useful purpose for the vast array of trees that cover the Earth's surface. The ultimate use, however, is one that calls that extensive coverage itself into question. The subject of deforestation—the elimination of old, stable forests, to be replaced by cattle ranching lands in Brazil, or land developed for housing and industry in the United States—is driving much research these days. Forests are one of nature's balancing wheels against the rise of carbon dioxide in the atmosphere and the resulting global warming. Yet, at the very time that carbon dioxide levels are increasing fastest, so is deforestation.

There are some 40,000 forestry and related conservation workers (nonscientists) in the United States, and another 27,000 forestry and conservation scientists, according to federal statistics; the growth rate for the near future is rather low—8 to 10.7 percent over the decade. But increasingly, the issues of land use, soil preservation, parks administration, and tree-dependent industries such as paper and lumber will call on the forestry professions to provide guidance to industry and government.

About two-thirds of forestry professionals work for the federal, state, and local governments; the remainder are in private industry, usually at paper and lumber companies. Foresters also figure in recreation, land planning and preservation, agriculture, and environmental research.

Some 50 schools provide a bachelor's degree accredited by the American Society of Foresters. An increasing trend in recent years is to require forestry professionals to become more familiar with other parts of the forest ecosystems—the wildlife, plant life other than trees, even the microbiology. In addition, with the increasing regulatory requirements, a knowledge of forestry policy and land-use rules are vital.

Fund Raiser

Fund raising is a specialized skill in that there is no way to study it in college, nor is there a "National Association of Fund Raisers" that offers professional identity. Nevertheless, it is a critical component of the functioning of nonprofit organizations. The job requires a knowledge of the public-issues realm in which the nonprofit organization functions, marketing; it also requires outstanding sales and communications skills and a gritty determination to continue in the face of resistance. Offering an intangible service in return for money is a hard sell.

Fund raising is generally learned through experience and, for those who are inclined to this line of work, experience is often readily obtainable through volunteer internships at the organizations that could eventually become one's employer. Nonprofit organizations vary according to size. The large ones, with tens of thousands of members, require sophisticated direct-mail, television, and other marketing techniques. Small organizations, such as those representing a group of similar manufacturers or local environmental-activist groups, call more for face-to-face selling skills. In turn, there are "pitches" made designed to appeal to specific audiences—one type to the publicly minded individual who can spare $25 or $50 for membership and a magazine subscription, perhaps; and another to the wealthy who can contribute thousands or more for a particular goal.

The lack of a preferred academic or professional training opens fund raising to individuals with practically any background. Training in environmental issues, through science courses or public health and public administration, provides an initial boost. After that, one's career progress is dependent on one's financial success. Depending on the organization, and the individual, fund raising often can be a route to the upper management of a nonprofit organization.

Geological Engineer/Hydrogeologist/ Hydrologist

This more technological (rather than scientific) look at the Earth is benefitting from the new emphasis on groundwater and Superfund cleanups. The concentration on underground physical conditions makes the profession something of a specialty, but then, what part of America doesn't need or use water from the ground? Hydrology, specifically, tends to focus on water resources management (thus including above-ground and below-ground water) in support of agriculture and other natural-resource applications.

The growing problems in various parts of the country in obtaining adequate water supplies should make some states (in the West, in particular) more active in hiring these types of specialists. A decline in enrollments over the past decade should make finding a job easier now. Starting salaries are around $28,000 per year; oil companies pay considerably more.

Grass-Roots Coordinator

No, this isn't a scientist who attempts to grow greener grass. This title usually refers to an employee at national environmentalist organizations who provides guidance and resources to local groups in their efforts to control environmental actions in their locale. Conversely, when the national organization sounds a theme that it would like to echo throughout the country (an occurrence likely to happen around election time), this coordinator gets the word out.

Grass-roots coordination is one of the rungs up the ladder to management of national, citizen-funded nonprofit organizations. By working with local leaders, understanding their pressing issues, and translating that knowledge into action plans for the national organization, the coordinator becomes deeply involved in overall policy making. Grass-roots coordination is not taught at any school, but the job usually presupposes a knowledge of science, politics, and organizational psychology. Communication skills are paramount.

Salaries tend to be minimal for such work, and many grass-roots coordinators are doing it on a part-time basis. On the other hand, the work provides excellent training in how environmental issues and social concerns impact each other on a local level.

Hazardous Materials Specialist

The transport, use, and disposal of hazardous materials is one of the front lines of environmental work. This job title, which has come into more common use because of federal environmental and workplace-safety laws, originally showed up only at chemical or oil companies. Now, it is common at pharmaceutical firms, hospitals, sanitation departments, fire departments, government environmental agencies, and shipping companies. There are plenty of these jobs at consulting firms as well.

The hazardous materials specialist (colloquially, the "hazmat" or "haz-ops"—for hazardous operations—specialist) works closely with industrial hygienists and safety engineers (see entries); the latter two professionals set the policies that assure a safe workplace environment, and the hazmat specialist carries them out. These duties include monitoring worker operations, inspecting storage areas or transport facilities, and seeing to it that the appropriate paperwork has been filed upon shipment or delivery of hazardous materials.

In hospitals, pharmaceutical firms, laboratories, and health-care facilities, the job title is sometimes "biohazards specialist." The functions are the same, except that a knowledge of biological hazards (infectious materials, radioactive compounds, and so forth) is required instead of chemical knowledge.

A related specialty, which usually goes by the title of "emergency response specialist," involves professionals who handle fires, chemical spills, oil spills, and other disasters. When a fire department, for example, races to a warehouse fire, the safety specialists need to find out rapidly what chemicals or materials are present, and how they may be best contained while the fire is smothered.

As these varied examples show, many different educational or experiential backgrounds are called on for hazmat specialists. One can study for a bachelor's degree in safety engineering, industrial hygiene, or health technology. Alternatively, a general background in industrial work can be combined with certificate courses in hazmat operations (see Appendix E). Salaries will vary widely, depending on the level of responsibility, the type of employer, and the specialized knowledge or experience one has.

Industrial Hygienist

The Stealth bomber that figured in the recent Iraq-Kuwait war was produced by an aerospace contractor using the latest technology in polymeric composite materials. The wings and other body parts aren't made from aluminum or other metal, but rather molded from liquid plastic resins, fiberglass, and other materials. Workers who constructed the planes complained of a host of mysterious maladies during the construction job. Are these illnesses job related? Are they related to the use of the new composites? Is there a way to protect against them if they are?

These are the sorts of issues that industrial hygienists address. In a time when new manufacturing techniques are being tried throughout industry, new safety standards must be instituted as well. Industrial hygienists are versed in human physiology and learn techniques to analyze worker operations to determine where problems can occur. In turn, these workplace standards set the stage for all the usage instructions that consumers receive when they purchase paints, household cleaners, tools, and other products.

Most recently, there has evolved a set of standards for carrying out environmental work, such as removing asbestos from buildings or cleaning up toxic dumps and spills. When thousands of workers were hired by Exxon Co. to clean up the Exxon Valdez spill, industrial hygienists went along to attempt to minimize health risks to those workers.

Undergraduate programs in industrial hygiene are rare. Most practitioners study a subject such as chemistry, biology, or engineering, then add on graduate courses in industrial hygiene. Starting salaries are above $24,000 and can rise quickly based on experience and employer. There are jobs among federal or state agencies (in particular, the Occupational Safety and Health Administration—see Appendix C), at large manufacturing companies, and among consulting firms.

Interpreter

Interpretation is the technical term for guiding or teaching visitors in the biosystems and natural history of the park or wildlife refuge they are visiting. The job involves the skills of a teacher and an awareness of what safety and park-use rules should be followed when, for example, a walking tour is being conducted. Interpretive jobs are very often seasonal, and the pay is minimal. In favor of the job, however, is the valuable skills it teaches those who plan to go on to develop more professional careers.

Being an interpreter is a good job for college students, because the hiring season often corresponds with summer vacation. When looking for employers, be aware of opportunities available locally or at state or private parks—not just the grand federal parks in the West.

Lawyer

Liability. Penalty. Criminal felony. These are some of the worries that keep corporate managers alert these days where the environment is concerned. Many federal and state laws passed during the 1980s included stiff penalties—including jail time—for business managers responsible for intentional pollution. In late 1990, according to a boastful U.S. Justice Department, indictments jumped 33 percent over the preceding year, conviction rates were running at 96 percent, and fines totaling $56 million had been assessed.

These legal concerns represent the ultimate outcome of pollution regulations. Most industry management tries to comply, and one of the main resources for developing that compliance record is the advice of an environmental lawyer. The latest laws are especially complex, and the interpretation of their regulations is an onerous task for industry managers. It is no surprise, then, that environmental law is the fastest-growing practice at law firms.

In the legal profession, the choices are generally among solo practice, joining a law firm, or working as a corporate lawyer. Not to be overlooked are government positions, through which many lawyers obtain highly valued experience in environmental regulation. While the big-time, large law firms, with 250 or more lawyers, can offer salaries of $75,000 and up to their new hires, the average income of all lawyers is actually considerably lower—

around $35,000 for those starting out, rising to around $100,000 for the experienced.

An interesting facet of environmental work is the evolution of new types of regulations and enforcement procedures. Can one legislate recycling of postconsumer wastes? Can pollution "rights" be traded, much as stocks are bought and sold, in a market-driven effort to determine the industries where pollution can be controlled most efficiently? How can a debt-for-nature swap be organized to help an indebted Third World country preserve its forests? Many lawyers find themselves involved in recommending how best to bring all this about.

"To be a good environmental lawyer, first you have to be a good lawyer," comments Clyde Szuch, managing partner at Pitney, Hardin, Kipp, and Szuch, a Florham Park, New Jersey, law firm. "An undergraduate technical degree would certainly be useful for the budding environmental lawyer, but it is not necessary."

Lobbyist

Lobbying is an insider's game. Lobbyists gain access to leading political figures and, by making a case for a supposed environmental improvement or preservation, effect change in government policy. Lobbying is traditionally thought of as occurring in Washington, but it is a practice in state and local governments as well. Lobbyists should have excellent speaking and communication skills and be able to think fast on their feet.

Lobbyists on the nonprofit side or on the industrial side must know law well and, therefore, many of them have legal training. Political science, urban planning, and economics are possible undergraduate programs. It is unusual for a lobbyist not to come out of years of experience in either political campaigning, representing industry in trade associations, or in grass-roots and national coordination of environmentalist campaigns. Many national associations are forbidden from formal lobbying by tax-code rules; lobbying organizations are in a different category from those that do not perform it.

In terms of pay, lobbying is one profession where there is a dramatic difference between industry-sponsored associations and activist ones. The former tend to have highly compensated, experienced lobbyists; the latter tend to have minimally paid practitioners. There are, of course, exceptions. Many industry-sponsored organizations (or even individual companies) hire professional lobbyists, paying princely sums for the access that these practitioners say they offer.

Noise-Control Specialist

Among the many programs that Ronald Reagan (well-known hearing-aid wearer) shut down during the early 1980s was EPA's research in noise control and the regulation of noise pollution. The business has by no means disappeared; all airports, for example, need to keep tabs on the noise levels around their facilities, as do many heavy-industry manufacturers. Acoustical engineering is a thriving field due to the growth of recorded-music media and advancing electronics. Some of this work is rubbing off in the noise-control arena, where, for example, manufacturers are developing methods of neutralizing noise by countering it with precisely offset sound frequencies. Noise control is of most concern to developers of heavy industrial machinery, where Occupational Safety and Health Administration regulations are still fairly rigorous. Noise control is also becoming more important to building dwellers.

Most research is the province of the advanced-degree holder. Starting salaries for Ph.D.-level scientists are in the range of $40,000–50,000; at lower degree levels, most engineers or scientists have salaries in the range of $28,000–35,000.

Planner

Everyone makes plans, but only planners make them for a living. Planning is a critical part of local and state government, as these agencies seek to guide the development of their economies, public services, and resource development on a rational basis. This rationality continually butts up against politics, history, tradition, and the individual desires of residents, constituents, and, not incidentally, taxpayers and voters. Planning can be a frustrating line of work, but it continually holds the promise of providing a better life for a larger fraction of the population, when natural and manufactured resources are used efficiently.

Planners accumulate information on current uses of public and private resources in a region, compare them with the desired future goals or needs, and try to develop a logical way of reaching those goals. Zoning regulations, for example, are a powerful tool for controlling the commercial, residential, and preservationist needs of a community. A city planner may recommend changing or establishing a zoning regulation to, say, preserve a historical landmark district or to prevent the overdevelopment of a lakefront.

According to the American Planning Association, the leading professional organization of the field, about two out of three planners work for government, with a quarter of the entire profession employed at the city level. The remaining third work in private business, education, or nonprofit organizations. Most have a master's degree or are in the process of obtaining one. Median salaries are around $24,000 for the relatively inexperienced, while those with more than 10 years' experience earn above $45,000.

Currently, the straitened circumstances of many city and state budgets, especially among the most populous states, make employment prospects rather dim. However, the need for more and better planning continues to grow. Time and again, the success of transforming economic backwaters into "in" places for people and employers to relocate to, while preserving the natural advantages of a region, has been demonstrated by professional planning.

Risk Manager

Risk management brings together the technical concerns of safety engineers, toxicologists, industrial hygienists, and the business concerns of corporate financial officers and administrators. It is an evolving position at most corporations, and the responsibilities vary from firm to firm.

In general, though, most positions share an interest in the liability insurance needs of the corporation and compliance with governmental regulations. "Risk management" as an academic discipline is taught at graduate business schools as part of a specialization in insurance issues. Other sectors of society, such as government agencies, insurance companies themselves, and environmental consulting firms, also are looking for the input of risk managers to help them deal with accident prevention and, when accidents occur, an apportioning of costs.

Typically, a risk manager examines the policies offered by insurance firms and compares them with the costs of self-insurance (which is becoming more common among high-risk industries, due to the expense of purchased insurance). This manager also keeps apprised of new trends in governmental regulation, because as various activities come under such regulation, the responsibility for compliance carries with it a certain immunity to liability (provided that the regulations were fully met).

Thus, risk management combines the skills typically learned in an M.B.A. program with the specialized knowledge of the insurance industry. For this reason, an initial job in the insurance industry is often very helpful in getting a career going. Salaries depend on the type of employer and one's level of experience. For the top corporate jobs, sometimes reporting to the environmental manager or even to the board of directors, earnings can be in the neighborhood of $100,000.

There is also a considerable amount of research being conducted with the goal of finding better predictors of the ultimate consequences of what activities are condoned today. Specialists trained in actuary science, statistics, public administration, and other programs are leading these efforts. There is little question that the responsibilities of the risk manager of the future will be quite different from those that exist today.

Safety Engineer

Safety engineers work in conjunction with industrial hygienists, hazardous materials specialists (see separate entries), and other professionals to ensure a safe workplace. Many of the techniques developed by safety engineers carry over into the design of products that are used by everyday consumers; such products as tools, chemical compounds, heavy machinery, automobiles, and the like show the influence of safety engineering.

A very pronounced trend during the 1980s has been the development of so-called "human factors"; the term refers to the measurement of how individuals interact with machinery, or how they use everyday items like chairs or ladders. Human factors and a related term, ergonomics, are strongly influencing the design of office equipment, computer work stations, airplane cockpits, kitchen utensils, and many other products. Because discerning consumers now demand ergonomically designed products, manufacturers are looking closely at the designs of their merchandise and employing safety and human factors consultants to aid in improving those designs.

Safety engineers work at all major manufacturers, at public agencies, and for consulting firms. Many students combine an undergraduate degree in one or another engineering specialty with graduate schooling in safety engineering, or with the many courses offered for safety and industrial hygiene (see Appendix E). However, an engineering degree is not mandatory. Starting salaries for B.S.E. graduates are around $30,000.

Soil Scientist

Soil science is one of the agronomy (see entry) specialties that has traditionally been most closely tied to agriculture. The majority of job slots have been with government, mostly for the benefit of agriculture through the Soil Conservation Service and related agencies. In private industry, a number of soil scientists have been employed by agricultural-chemical manufacturers, especially fertilizer suppliers. In the future, while this agriculture-soil science relationship will continue to dominate in terms of job slots, other sectors of the U.S. economy are calling for the expertise of soil scientists.

In particular, environmental consulting companies that perform Superfund-related cleanup work or that are involved in correcting or preventing the pollution of underground water reservoirs are hiring soil scientists. Landfills, much bemoaned by environmental activists and by city administrators, will remain a dominant factor in solid-waste disposal for years to come; the planning and operation of these facilities require soil science expertise. And, as land-use issues rise in importance in both populated and wilderness areas of the country, soil scientists are called on to help analyze and plan for environmental impacts.

Only a few hundred students study soil science in college these days; along with a slump in enrollments for nearly all agriculturally related majors, soil science has fallen as well. In recent years, there have been almost as many graduates with doctoral degrees as with bachelor degrees—an indication of the research orientation of the profession. Federal salaries range from about $16,000 to $28,000, based on the degree held. Private industry is usually 10 to 20 percent higher.

Technical Writer

This title is being used to lump together a variety of positions that require skill in converting records or oral reports to finished documents. These positions share many of the same responsibilities as that of the community relations manager (see entry), with the proviso that, usually, the communications are between one organization and another rather than between an organization and the public.

A frequently seen title is "proposal writer." This individual has the responsibility for gathering together the technical and financial information that a consulting company develops in preparation for making a bid on a government or industry contract. The chances of winning these contracts are aided by delivering a well-written, polished proposal, and thus writers are hired to provide those finishing touches.

Once a project is concluded, a final report must often be issued, both to record what has been done and to provide useful information that might help in later projects. A job title sometimes seen in this context is that of "documentation specialist." Scientists conducting research are usually required to submit such reports when their studies are concluded, and some organizations employ editorial specialists to complete the documentation.

Yet another example can be seen among consulting firms or educational institutions that provide training for environmental specialists. Very often, there is no standard textbook for what is being taught; the instructor develops his or her own course material, which represents the collective experience of the organization offering the training. A writer can help prepare these materials.

A final example—one that I'm obviously fond of—is writing for technical publications or book publishers. These organizations often employ journalists but, in some cases, the job requires some level of technical training in addition to writing skills. With the right foresight, a journalism major can specialize in such technical writing by taking additional science or policy courses while in school or as a graduate student. Salaries vary considerably, depending on the type of employer and level of experience, but usually, the technical writer can count on earning slightly more than the typical newspaper journalist, which is currently around $19,000 for the recent graduate.

Toxicologist

Toxicology is the study of the adverse effects of the environment on human health. (There is also a subspecialty in animal toxicology.) The emphasis is usually on harmful chemicals; diseases caused by bacteria or other life forms are usually studied by biologists or epidemiologists.

Toxicologists spend lots of time in a laboratory, functioning essentially as detectives. A certain medical condition has been encountered; what chemical is the cause of it, and where did that chemical come from? Because of this close connection between chemicals and human health, many toxicologists work in the pharmaceutical industry, both to ensure the safety of pharmaceutical compounds and to preserve the health of workers. Toxicologists were among the first to discover the link between lung diseases and asbestos, the degenerative effects of lead, and the dangers of air pollutants such as ozone and nitrous oxides.

It is possible to study toxicology at the undergraduate level, but many practitioners go on for graduate degrees. Some also take the route of obtaining a medical doctor degree.

Besides the pharmaceutical industry, jobs are available at chemical companies, food and cosmetics manufacturers, public health agencies, consulting firms, and in academia. For doctoral-degree holders, current salaries are in the range of $36,000–45,000. Certification is available through a variety of professional organizations.

Water Quality Technologist

A decade or so ago, it was common in most parts of the country to take water totally for granted. Relative to many other countries in the world, the United States was blessed with plentiful supplies of clean water.

In reality, those water supplies were already the product of much technical expertise in developing water resources, transporting them to where they were needed, and arranging for their distribution. The system ran very well with little public attention.

Today, the situation is changing. New sources of contamination appear, causing water wells to be shut down or water treatment plants to go temporarily out of service. A drought—not the first, but the first with California as the most populous state in the country—in western U.S. regions is changing agricultural plans dramatically.

All these problems make the prospects for careers in water quality management brighter. This specialty, taught (usually at graduate levels) at many of the land-grant colleges of the Midwest and West, calls for familiarity with a diverse array of physical and social sciences: chemistry, geology, hydrogeology, public administration, and agronomy. The simple fact of the matter is that as population increases, more demands are put on all natural resources, water included.

Many water quality managers work for local, state, and federal government. Water quality managers oversee the operation of water and wastewater treatment plants, irrigation systems, and urban water-delivery systems. Other employers include agriculture and industries that are heavy consumers of water—pulp and paper manufacturing, public utilities, and food processing. Managers in these industries must ensure the quality of water going into the production machinery of the plant, as well as maintain control over the contamination of the water resulting from its industrial use. Manufacturers, and the engineering consulting firms that serve them, are also required to assist in cleaning up previous contamination of groundwater and preventing future contamination.

Salaries vary considerably based on the type of employer and level of education. For bachelor-degree holders, starting pay is around $18,000 in government employ and approximately 10 to 20 percent higher in private industry.

CHAPTER SIX

Entering the Job Market

One way of looking at a career—almost any career—is to imagine a long corridor filled with doors. As you enter through a door along this corridor, another hallway beckons with more doors along its walls. Making a career move, whether as an unemployed new graduate to the ranks of the employed, or as a midcareer strategic move from one line of work to another, is a step through one of those doors. With each passage, certain opportunities open up and others are foregone. There are any number of connecting corridors with only one other door in them; it is difficult, though, to imagine a corridor with only one door (the entrance) in it.

In common with most other professions, the environmental field has many doorways leading into it and many paths to take within its territory. The choice for a first job, or the choosing of a particular academic discipline, opens some doors; the jobs or forms of study not taken will present certain limitations.

Two very special conditions dramatically affect the vista of environmental careers today. The first is that the field is growing rapidly. More public and private funds are flowing into environmental work and will continue to do so for years to come. Service companies are growing from a handful of associates to 1,000 employees or more, in offices spread throughout the United States and abroad. The second special condition is the rapid pace of change the field is experiencing. There are whole job categories, with attendant professional training and practices, that barely existed five years ago.

This chapter will lay out some of the details of these trends and show how people are using their experience and training to win green-collar jobs. An important proviso to all this information is that the "standard" practices are evolving and changing, just as the types of work are changing. In the course of time spent writing this book, about a dozen new professional societies were formed; these will form a locus of job opportunities for future professionals. More than a dozen new schools of environmental education were organized as well, which will put a new stamp of expertise on their graduates. Both educators and professional societies, as well as government agencies, are giving serious consideration to new types of professional certification.

There are many simple, practical considerations to getting a green-collar job: where to find out about opportunities; what forms are to be filled out, what kinds of educational preparation are currently accepted, and what are the customary paths of advancement. Many of these details will be presented here, as well as in the appendixes at the end of the book.

Technical versus Nontechnical Jobs

Probably the most widespread differentiation in green-collar jobs is between those that require a technical background and those that do not. "Technical" can be taken to mean those jobs in which engineering or scientific work is done or most of the medical and public-health professions. "Nontechnical" can be taken to mean that the primary skills are communications and interpersonal relations, community activities, and public policy making. There are many jobs that combine technical and nontechnical elements, and many that are indeterminate between the two—a technically trained person could bring some skills to a job and an nontechnical person can do the same.

Obviously, a liberal arts major cannot do analyses of the chemistry of the upper atmosphere successfully. But can, say, a business graduate manage a firm where atmospheric scientists work? Or can a fine arts major run a community relations program in which a manufacturer's production plans are reviewed and approved by the community?

"Yes," says Mark Johnson, a project manager at PRC Environmental Management, Inc. (McLean, Virginia), who has the distinction not only of managing Superfund projects, but also, during earlier work assignments at PRC, of helping establish financial and technical requirements for RCRA-related regulations that the EPA established under that program. Mr. Johnson, whose undergraduate studies were in occupational therapy, later gained an M.B.A. in insurance and was soon deeply immersed in EPA-related work. "What I deal with day to day is multidisciplinary—chemistry, hydrogeology, civil engineering, and much more," he says. "When I get lost, I tap the technical resources in my company. If anything, the education I wish I had would be in law rather than in technology."

"Yes," says Stephanie Reith, a community relations manager for Donohue & Associates, Inc., a Chicago engineering, architecture, and research firm. "My education was in art criticism, and I came to the environmental field through a desire to do marketing work," she says. After a few years, her employer asked her to take over community relations responsibilities. "I've had to learn about earth science, engineering, chemistry—the whole lot. I use the tools of any good researcher: I ask questions, I do my homework, and I try to relate the extremely technical information I gather into terms that the public understands."

Are these two professionals the exception or the rule? To answer this question, the distinction must be made between workers in environmental businesses and the rest of the green-collar work force. In the environmental businesses—manufacturers, pollution-generating organizations, and business-service firms—the conclusion is that they are the rule now, but will become the exception in the future. The essence of environmental work is the life sciences, the physical sciences, engineering, and the health professions. The environmental business has grown so rapidly in recent years that it is drawing in professionals from all walks of life, as the preceding examples show. Environmental businesses have been forced by the limited supply of adequately trained workers to hire those with nontechnical backgrounds. In the future, these businesses will continue to hire nontechnical workers, but only for specifically nontechnical positions such as public relations, accounting, office administration, and the like. Many companies today pride themselves on the depth of their technical staff, and this will only increase as the competition among these firms intensifies.

In the nonbusiness arena, which includes all the nonprofit organizations, some types of government work, and some types of business consulting, the field has been wide open and will remain so. Many old-time activists in environmental organizations bemoan the increasing "professionalization" of their groups, having evolved from a collection of like-minded activists from all walks of life into smoothly running organizations with experienced fund raisers, money managers, or technically astute professionals. But by this professionalization those old-timers really mean the difference between someone whose commitment to the environment propelled them into the position of, say, an association manager, and the individual who studied association management and then joined an environmental group as its manager. They are not necessarily referring to individuals with a technical background. There will probably always be new organizations that form out of newly identified environmental needs and draw on the "general public" as members.

The traditional knock against a technical background in science or engineering for typical entrants into the job market is that they lack communication and interpersonal skills, which growing, manager-hungry firms desperately need. Scientists and engineers are the nerds, in this view. Bill McKibben, author of the pathbreaking recent book, *The End of Nature*, let this bias show in a column he wrote for the British newspaper, *The Independent*: "Last year, I spoke to the students and faculty of MIT [the Massachusetts Institute of Technology, which graduates primarily engineers and scientists] . . . of the class known in America as techno-dweebs." What useful knowledge this characterization conveyed to his British readers is hard to say. But this down-putting view is all too common, even among technical professionals themselves. The clear message to all this is that communication skills are especially important in the green-collar work force. This includes, obviously, writing as well as speaking.

Conversely, when you talk with a person without a technical background who happens to be doing highly technical environmental work, the statement is usually made early in the conversation, "Even though I don't have a technical education, I need to . . . ." This chip-on-the-shoulder frame of mind could have crippling disadvantages if it goes too far. But most green-collar workers today have realized that as long as they are willing to learn and to take the time to study on their own, most technical knowledge can be mastered or can be put in its relevant place when trying to resolve larger issues. This puts the burden on the nontechnical worker to keep up with evolving technology and technical issues. In reality, both the technically and nontechnically trained must spend a lot of time keeping up with new developments. To cite just one example, the science of the upper atmosphere detailed in Chapter 7 is vastly better known today than it was only a decade ago, and this knowledge will necessitate many changes in other areas of science and technology. But nontechnologists who contemplate a career in the green-collar work force must be ready to contend with a continual challenge to their expertise.

In today's job market, opportunities are plentiful enough that the uncommunicative, unsocial technologist, or the nontechnologist who breaks out in a rash whenever an equation is used, are both being hired. As a later section in this chapter will show, there are plenty of continuing education opportunities to make up for both failings. But, as the environmental field grows and evolves, and as more conventional college-level environmental programs open up, the level of expertise expected of new job seekers will rise.

Volunteer Work

One of the rather unusual aspects of environmental work, relative to many other professions, is the availability of internships, volunteer programs, work/ study programs, and other types of "experiential education." Volunteerism is an integral part of America's environmentalism for an obvious reason: We all live, work, and play in the environment, and if we care about it, we will get involved. "Think globally, act locally" has been the rallying cry of the environmental movement since the mid-1960s, and it remains truer today than ever before.

For those contemplating a green-collar career, these volunteer opportunities (some of which, in fact, provide some pay) offer great benefits—the chance to gain valuable experience, to learn new things, to see some of the issues that are driving the environmental movement firsthand, and, not least, to demonstrate to a potential employer your interest in and familiarity with environmental issues. And, let's face it, if you are sincere about your interest in the environment, the question to ask is not, "Should I get involved in volunteer work?" but "Why aren't I already involved?"

First, some definitions. A volunteer position is just that—you ask for it and, within the constraints of the organization, its needs, and its current staffing, you fill a role. You have a great degree of freedom in what to get involved in simply because your labor is freely given. You can choose among organizations and among functions within an organization. The student or professional looking ahead to a green-collar career should be able to accumulate experience, and, at least, get a reference letter out of the work.

Internships, by contrast, are usually for a set period of time and for a more defined purpose. They may or may not be paid positions. In return for fulfilling the requirements of the internship, the intern should expect to carry something away from the experience: a report that was completed; a set of skills that was demonstrated; and, of course, references and maybe leads to an actual job. Some internships are sponsored (paid for) by one organization, but the actual work is carried out at another, an arrangement something like a privately funded scholarship that pays for classes at a college.

Speaking of colleges, there are a great number of opportunities for internships or volunteer jobs that are spawned either at, or in close association with, higher education. These range from lab-assistant positions to helping an economics or political science professor, say, conduct research. In some cases, the line between a volunteer job and actual course work blurs; you can turn volunteer experience into credit toward a college degree.

The next step closer to a real job is a work/study program, which is also known as cooperative education. Some employers and some neighboring schools have arrangements whereby the student spends five to six years earning a baccalaureate degree and works at a full-time job during summers and part of some academic years. Only a few schools offer such formal programs, although at large universities (where the same course might be offered both in spring and in fall), an enterprising student could "construct" his or her own work/study program, as long as there is a suitable employer available.

Volunteerism is a great way to gain valuable insights and experience. The fact that his or her work is being freely given puts the job seeker in the driver's seat, able to pick and choose among activities and organizations. For the already experienced or the already working, a volunteer position offers a means of broadening their background and a way to network with other professionals. For the young job seeker who considers a volunteer position the best or only avenue toward a desired career, a word of caution is in order. Don't let the relationship degrade into one where you can be taken advantage of, compelled to work longer hours at tasks less helpful to your career, on the oral promise that you will be "taken care of" when the work finally ends.

The Complete Guide to Environmental Careers, written by members of the CEIP Fund (which formerly went by the name of The Center for Environmental Intern Programs), offers many insights into volunteer and intern programs. The Fund has placed nearly 4,000 college students and recent graduates in internships, leading to its claim to being "largest on-the-job trainer for environmental careers in the country."

The National Wildlife Federation (see Appendix A for address) publishes an annual conservation directory that lists hundreds of local and national environmental organizations, as well as some details on volunteer or intern positions. The federation itself sponsors numerous internships.

The federal government and some state government agencies sponsor many internships. Write to the agencies listed in Appendix C and D, or look up a new publication called *Career America: The Federal Career Directory* at your library or job-placement center.

There are also numerous directories and newsletters of volunteer positions and organizations. Check with your local library for details.

Government Service

There are over 3 million civilian employees of the federal government, over 4 million state employees, and over 10 million local-government employees.

The federal government is the single largest employer in the United States. And, according to the U.S. Office of Personnel Management (OPM), the government hires more than 300,000 new employees every year. Thousands of those hirees are green-collar workers.

In Appendix C, I've listed names and addresses of all the major departments and agencies of the federal government that have a direct relation to environmental protection and preservation. (Some of these agencies have also been accused of being the worst destroyers of the environment as well, but that debate goes way beyond the scope of this book.)

This list includes the obvious: the U.S. Environmental Protection Agency, the Department of the Interior (including the National Parks Service and others), and the Department of Agriculture (including the Forest Service). It also includes the Bureau of the Census, National Institute of Standards and Technology, National Oceanic and Atmospheric Administration, the Army Corps of Engineers, Department of Energy, Public Health Service, Occupational Safety and Health Administration, National Aeronautics and Space Administration, National Science Foundation, and Nuclear Regulatory Commission.

Federal agencies that have many more applicants than jobs generally let OPM serve as the initial interface between job seekers and the target agency. OPM maintains Federal Job Information Centers in most (but not all) states. It also coordinates job postings with the State Employment Service of the Department of Labor; this service is available in 2,000 cities nationwide (look in your local phone book). Finally, it maintains a Career America College Hotline for information seekers: Dial 1-900-990-9200. (Please note, this call will cost you some money, but only 40 cents per minute.) The federal government has special programs to aid in hiring minorities and the physically disabled.

Assuming that a vacancy to your liking has been located, you will usually be required to complete all of the following forms: the Standard Form 171 (SF 171); a résumé-like pamphlet; OPM Form 1170/17, which allows for a listing of college courses; and a résumé. Some positions require passing an examination, the "Administrative Career with America" exam; but some positions are exempted.

Often, it is enough to show what qualifications you have and how good your grades were in college to be accepted. The federal government hires competitively, but the overall system works more in the direction of demonstrating the best existing credentials, rather than subjective evaluations derived from interviews.

In Appendix E, there is a listing of state-level environmental agencies or those whose work is closely related to the environment. Please note that, just as the federal government has parks services, natural-resource management branches, and the like, so do most states; listing all of them here would be a monumental task. The same is true for local government.

Generally speaking, green-collar employees in government work either as administrators, overseeing the work of private contractors, or as regulators, examining the records and applications of private industry and making a ruling. Only in particular jobs—research and emergency response come to mind—does the government worker get to run things with his or her own hands.

There are two interrelated career-building aspects to government work. One is that the young worker is thrust into major responsibility early on. Of course, the majority of actions are subject to continual review, but even so, the amount of responsibility taken on is huge compared to a private-sector job. The drawback to this exciting opportunity is that there never seems to be enough resources to go around; staff is short-handed (that's why the responsibility comes early—there's no one else to give it to), and there are difficulties in getting decisions made.

The second aspect, mentioned in Chapter 4, is that government work is an excellent preparation for a private-sector job.

Private-Sector Jobs

The routes to a job in the private sector are almost as varied as the number of employers. In certain sectors (engineering consulting is one), employers seem so desperate that they are going door to door to get properly trained new employees. Companies are raiding each other's staffs through the use of executive-search firms, and government employees are siphoned off readily.

This intense demand is not consistent across the board, and it probably won't last much longer. Many engineering companies, for example, jumped into environmental work when construction projects for private industry slumped in the early 1980s. As construction came roaring back in the late 1980s, shortages began to appear. But now, with a recession in place as of early 1991, there may be a corresponding swing of engineering staffs back to environmental projects.

For a select group of green-collar workers, the upsurge in environmental interest will probably mean many opportunities for years to come. This group is composed of the researchers and inventors who develop new technology. Both for remediation work (Superfund, in particular) and for the development of green products, a researcher who develops a new technique is suddenly the toast of the town. For example, a scientist who devises a way to remove a toxic compound from soil will find his or her phone ringing with offers from venture-capital firms that want to build a company around the technology, commercialize it, and then "go public" (i.e., seek a listing on a stock exchange). The process has been well worked out through the experiences of Silicon Valley microelectronics firms in the 1970s and the biotechnology industry that came to the fore in the 1980s.

There are some $2 to $3 billion available for investment at any given time, according to venture capitalists speaking at ETEX 91, a new environmental business conference that featured several representatives of the financial community. However, these venture capitalists were quick to point out that they do not want to take an idea from an inventor's mind and spend the money necessary to prove the idea has value. Rather, they prefer to enter the scene when the inventor has already proven the technique in a laboratory or test demonstration. Proving out an idea can cost millions of dollars, and for this the inventor usually depends on a group of "angels"—family, friends, and other personal resources. It has been estimated that twice as much money is committed to new technology through angels than through venture-capital firms. The message is, then, to be ready to develop your own business-management skills and to uncover your own sources of capital before the venture capitalists appear on the scene.

By no means should you believe that only a high-tech scientific inventor can develop a new company or solve a pressing environmental problem. The biggest type of new environmental business in early 1991 is probably individuals setting up recycling ventures to recover a valuable material from industrial or municipal waste. Often, this is no more technical than finding an empty building where the recycling can be done and spending countless hours coordinating with local communities for the supply of material to recycle.

By far the biggest grouping of private-sector employees are those who work at manufacturing firms, which have been increasing their internal staffs for health, safety, and environmental affairs and committing more funds to new product development. These employers want job candidates who can do everything—who have specialized technical knowledge, yet can apply multidisciplinary skills to any problem; who are familiar with machines and technology, yet have great communication and interpersonal skills; who have vaunting ambition to move into management, yet who are willing to be hired into low-level jobs and do what they are told.

There are probably a few of these Supermen and Superwomen around, if only to make the rest of us feel humble. Rather than masquerading as one of these otherworldly beings, the more realistic attitude to have when seeking private-sector jobs is to continue building your experience and skill set while in school and while in whatever occupation you are in currently. If your knowledge of technology is good, but your awareness of environmental regulations is not, seek extra training. If you seek a marketing role, but lack the necessary sales experience, look for ways to get involved in sales and marketing. Always work on your communication skills, no matter how confident you are of them.

How are the private-sector jobs uncovered and how can you get them? One of the clichés of the recruitment business is that for every job advertised in the newspaper, there are 7 or 15 or 20 unlisted jobs available. (Various surveys of this phenomenon have come up with various counts.) That statistic, however high it is, points to the need to network. Join a professional society—either its student chapter or the main society—as soon as employment starts. Go to meetings and get involved in the management of the local sections or in setting up national meetings. (Like government, there is always more work to be done in a volunteer professional society than there are hands to do it.) Many of these organizations are listed in Appendix A.

Education

The environmental business at the dawn of the 1990s looks very similar to the computer and data-processing industry of the late 1970s. Suddenly, new interest and/or new technology creates intense demand for highly skilled professionals. But there are not enough of these professionals to go around, so people with all kinds of prior education or work experience are hired and learn what they need to know on the job.

There is a substantial difference in level of technology between environmental work, taken as a whole, and the computer industry, which is highly specialized and very focused technically. But the similarities are compelling. Following the upsurge in computing activity during the 1980s, students flocked to computer science and electrical engineering programs. When the industry slowed down around the middle of the decade, the job market suddenly seemed inhospitable, and many students opted out of computer science. In fact, employment continues to grow. The students who dropped out were primarily those who thought to enter the field as a way to make big bucks, rather than out of a personal interest in the field.

As those enrollments surged, the level of professionalism in the field grew. Jobs that didn't require a college degree gradually came to require it;

jobs for which a bachelor's degree was adequate now went preferentially to the master's graduate. You can expect some of the same things to occur in the green-collar work force over the next five years.

As things stand now, there is a wide variety of academic programs that can prepare you for green-collar work. Technical degrees are in higher demand than those in the liberal arts:

Life Sciences

Agricultural science (plant and animal), Forestry, Biological sciences (biology, biochemistry, botany, microbiology, ecology, zoology)

Physical Sciences

Chemistry, Geology and other earth sciences, Mathematics, Physics, Computer science

Engineering and Technology

Aerospace engineering, Architecture and architectural technology, Agricultural engineering, Chemical engineering, Civil engineering and construction technology, Electrical/electronics engineering, Industrial engineering, Instrumentation engineering, Manufacturing engineering, Mechanical engineering, Petroleum engineering, Surveying and mapping science.

The nontechnical lines of work employ graduates with practically any college degree (or indeed, those who lack of a college degree); however, there are many types of jobs that require a degree in, or familiarity with, the following:

Accounting
Agribusiness
Business and business management
City, urban, or regional planning
Communications (journalism, public relations, communications technology, speech)
Criminal justice
Economics
Education
Geography
History
Home economics (family services, consumer affairs, nutrition)

Law
Liberal arts (English, foreign languages, literature)
Library science
Parks and recreation
Philosophy
Political science
Psychology
Public affairs (public administration, public policy, social work)
Sociology
Visual and performing arts

As is readily seen, most of what is listed here is commonly available at colleges and universities across the country.

Assuming that you are a college student in one of these programs, or that you are a young professional with an already established academic background with work experience piling up, what studies should you add on? The programs that seem to be in highest demand currently are those that offer either a master's degree or a nondegree certificate in environmental management. These programs combine a smattering of science (to learn the basics of earth science and chemistry) with a thorough grounding in environmental law and policy.

Undergraduate studies in environmental issues, by and large, are being worked into existing curricula. Civil and chemical engineering programs, in particular, are rapidly adding new course work. Many schools are setting up cross-disciplinary or multidisciplinary programs that combine technology with public policy, business management, or economics. *The Right College*, a directory published by Arco Press, lists 55 undergraduate programs in environmental engineering, environmental studies, or energy conservation, but this is only the tip of the iceberg. Dozens more schools make environmental studies available as an option within a larger academic department.

At the graduate and continuing education level, there are hundreds of programs, both on college campuses and at private training companies and in government. Appendix E lists a good number of these, but more are being added almost monthly. The nondegree certificate programs are worth highlighting because they are able to fill a crucial gap quickly—the gap between what today's working professionals have and what they need in order to advance in their jobs, as well as the gap between what colleges will have available in future years and what today's students want now.

The Moral Dimension

One other special condition, which stands out relative to many other professions, affects the individual considering an environmental career. This is the moral dimension to environmental work. Many careers are defined mainly by their stability and the financial rewards they offer, and many of the workers in these careers are in it mostly for the money. That's not a negative, for most people. Earning a living is an essential fact for most of us, and the simple act of going to work and performing a useful function brings harmony to the lives of most of us. Some people contemplating an environmental career admit to these essential interests, but seek to perform work that preserves and protects the Earth as an expression of their religious and philosophical convictions. The philosophy of Earth First! is perhaps the most extreme, but clearest expression, of this desire: "In *any* decision, consideration for the health of the Earth must come first," writes Dave Foreman, one of the founders of that group, in *Confessions of an Eco-Warrior*, adding explicitly that the Earth must be placed even ahead of human welfare if necessary.

Those who work in public social services, charities, or religious orders speak of a "calling" to their work, which signifies that considerations such as salary, job satisfaction, stability, or advancement are secondary to their moral vision. Many entering environmental professions feel the same way. However, this moral vision is not an essential prerequisite to being part of the green-collar work force. To have it is an advantage in nearly all types of green-collar work, but to lack it, or more accurately, to believe in putting the health of the Earth in the context of human needs and activities is not a barrier to most types of green-collar work.

There are two reasons for asserting this—the practical, and the political. The practical reason states simply that it is clearly possible to do enormously beneficial work for the good of the Earth without having any special moral vision of the Earth. A scientist who develops a dramatically more energy-efficient light bulb, which conserves vast amounts of energy and obviates any number of pollution-generating power plants or hydroelectric dams, has done at least as much good for the environment as the activist who lies down in the road to prevent the construction of those power plants or dams. That scientist may be motivated by nothing more than the desire to make a buck by providing a clearly better piece of technology. Similarly, the civil engineer who brings bulldozers, chemical detergents, and processing equipment to a Superfund site to clean it up may be concerned with no more than meeting the technical performance specifications his or her contract requires and leaves the greater issues of environmental quality or the long-term health of the Earth to another forum. There are green-collar workers on the front lines of protest marches to save the Earth, and there are also green-collar workers spending endless hours in laboratories or government regulators' offices hashing out the day-to-day steps by which society advances.

The political reason involves an assertion I made at the beginning of this book, in which I stated that it is only when environmental protection and preservation have been worked into the warp and woof of American capitalistic enterprise can we be assured that the environmental movement is here to stay. This assertion is a controversial one; there are still many people in all sectors of society who believe that the current upsurge of interest in the environment is but another passing fad, an economic luxury that will continue as long as the economy can afford it. I maintain that the current intense interest in environmental protection and preservation is driven by the economy itself. Manufacturers are finding that they can produce more efficiently and more profitably when they don't pollute. Many sectors of the economy will now fight for preservation and protection out of their own economic self-interest. These include the entire travel and tourism industry, the insurance industry, and the health-care establishment, among others.

CHAPTER SEVEN

A Case Study: CFCs and the Ozone Crisis

Introduction

It is a major contention of this book that there now exists an identifiable sector in the U.S. workforce: the green-collar workers. They are employed in industry, in government, in nonprofit groups, and in academia. They are by no means united in their outlook toward the future or even toward society. But they share an expertise and a body of knowledge, and they are engaged in one of the most profound issues affecting society today: the health and well-being of the environment.

A case study can be used to show how these various elements interrelate. It can also show how the complexities of today's environmental issues involve many different elements of the American community. The case study I have selected—CFCs and the depletion of ozone in the Earth's upper atmosphere—is a story that has been told more than a few times. But, to my knowledge, it has never been told from the perspective of the workers involved in it.

Another element of the CFC story is that it is a good one, with as much adventure as a spy thriller, with good guys and bad guys, and with a resolution. Although, as you will see, the story isn't completely over (the repercussions will last for at least the next 25 years), the major elements have already occurred. An environmental problem was postulated, research was done to identify it, laws were passed to address it, and society is in the process of changing to eliminate it. I would go so far as to say that it has a happy ending, although there are many who would say that the entire episode represents an abysmal failure of twentieth-century society. However, both proponents of the process and detractors of it would agree that the controversy is shaping the future course of environmentalism throughout the world.

Refrigeration Worries

The story begins in the Dayton Engineering Laboratory in Ohio, which was owned by General Motors Corp. The laboratory was founded by Charles Kettering, one of the giants of automotive research in its early years (famous for, among other things, devising an automatic starter to replace the hand crank for starting a car). The laboratory was a hotbed of innovation modeled after Edison's laboratory in Menlo Park, New Jersey; Edison himself was at the very end of his career. Thomas Midgley, a Ph.D. mechanical engineer from Cornell University, was employed at Dayton as a researcher. (See B.A. Nagengast, "A History of Refrigerants," and R. Downing, "Development of Chlorofluorocarbon Refrigerants," in *CFCs: Time of Transition.*) In 1928, Kettering told Midgley that "we have come to the conclusion that the refrigeration industry needs a new refrigerant if it ever expects to get anywhere." Three days later, Midgley had developed and had begun testing the first chlorofluorocarbon refrigerants.

Today, the refrigeration and air-conditioning industry is known collectively as "HVAC" (for heating, ventilating, and air-conditioning). The reason the HVAC industry of the 1920s "wasn't getting anywhere" was due to the danger and inefficiency of refrigerants at the time. A refrigerator needs a fluid that can be compressed and cooled, then allowed to expand into a chamber. The expansion causes the fluid's temperature to drop rapidly, and this coldness provides the refrigeration. Refrigerators at the time used ammonia, carbon dioxide, sulfur dioxide, and methyl chloride as refrigerants. Some of these are useful refrigerants; ammonia is still used in large commercial systems. But when manufacturers attempted to make small refrigerators for the home kitchen, they ran into trouble. Ammonia and methyl chloride are toxic and could (and did) kill people when they leaked. Methyl chloride is also flammable, and sulfur dioxide forms sulfuric acid—one of the most powerful acids—when in the presence of water.

Midgley was able to quickly identify candidate materials by studying the chemical properties of these conventional refrigerants and theorizing which compounds would have the best combination of refrigerative and safety properties. He selected simple hydrocarbons that could react with chlorine and fluorine. Safety was on his mind: some of the reactants could form phosgene gas, the essential component of the war weapon mustard gas, which is fatal upon inhalation. With a doctor present, the gases were tested by exposing a guinea pig to them. Very quickly, the scientists formed the compounds that are now known as refrigerants R-11, R-12, and R-21 (which correspond to the more modern "chlorofluorocarbon" designations CFC-11, CFC-12, and CFC-21).

In 1930, Midgley unveiled the new compounds at the national meeting of the American Chemical Society. The highlight of the presentation had Midgley inhaling a lungful of the gas and extinguishing a lighted candle, demonstrating (one assumes) both its nontoxicity and its nonflammability. (Stunts like this were what constituted toxicity testing in those days.) At some point, the CFC research left Midgley's laboratories and entered those of Du Pont, where the process of "scaling up" the production of the material was worked out by engineers and chemists. Plants were built and production steadily rose. When the patents ran out in the early 1950s, other companies in the United States and abroad began manufacturing the compounds without paying a licensing fee to Du Pont.

New Markets

The success of the CFCs led to the formation of a new company, Kinetic Chemicals, jointly owned by General Motors and the Du Pont Co. Du Pont bought out GM in 1949; along the way, the refrigerants came to be known as Freon. Manufacturing of the material continued rapidly through the 1930s and 1940s, and by 1950 Freon was well established as the refrigerant of choice for most components of the HVAC industry. The key property that led to its success, besides its heat-absorbing properties, was its near-absolute inertness in refrigeration applications. It didn't break down, and, by and large, it didn't react with or affect the mechanical parts it touched. You could put it inside the refrigeration coil and basically forget about it for the life of the unit.

During the 1950s, too, Freon was investigated for other industrial applications. Its inertness to other chemicals, combined with its ability to "solubilize" (i.e., dissolve into) other chemicals, made it a good candidate for propellants in aerosol spray cans. By pressurizing the material when filling the can, a material such as paint or hairspray can be sprayed outward in a smooth, even coat. The nonflammability of the material made it safer than other propellants such as hydrocarbon gases. This dissolving power also made it ideal for industrial solvents—materials that clean metal surfaces or other manufactured parts while they are being assembled or repaired. Meanwhile, the mass consumerization of America continued apace. Freon helped air-condition the South and Southwest and made automotive air conditioners less expensive and more reliable. During the 1950s, and much more so in the 1960s, Freon helped make polyurethane foam a popular material for furniture padding and similar applications. This ubiquitous foam is made more uniform and a better insulator when produced with Freon, relative to other production techniques. (In this application, Freon acts as a "blowing agent," forcing the foamy plastic through a nozzle and into a mold, where it solidifies in its characteristic form.) A final major application, to become very important in the late 1960s and 1970s, was as a cleaning agent for electronic subassemblies. Everything from the innards of television sets to high-tech computers are manufactured with CFCs as a cleaning agent during intermediate steps.

By 1970, worldwide production of CFCs was reaching two billion pounds per year. Where was it all going? To those who wondered about the subject (and there were a few), the assumption was that they drifted into the atmosphere, which was considered an "infinite sink." This phrase refers to the belief that when a relatively small volume of material is dropped into a enormously large space, it literally vanishes—its concentration becomes too small to be measured. It was assumed that there was a natural breakdown of the material by sunlight, cosmic rays, or whatever; maybe some of it drifts out to space. (The Earth's oceans used to be considered to be an infinite sink as well; in some circles, they still are believed to function this way.)

In the late 1960s and early 1970s, a dispute arose over whether or not the U.S. government should support research into developing a supersonic transport (SST). This jet, by flying higher and faster than the conventional passenger jets of the time, would revitalize the aircraft industry and provide better service to the international traveler. American business leaders were also worried about European competitors, whose aerospace firms had already organized into a collaborative effort. By now, the many environmental concerns of all forms of transportation were being reexamined by scientists and environmentalists. Anyone looking into the sky could see the very unnatural contrails formed by conventional jets as they flew; what would the effects of such contrails be higher in the atmosphere? One obvious threat could be from nitrous oxide compounds, which form nitric acid in the atmosphere. What would this do?

The SST fight in the United States was a nasty one, made more embittered by the Vietnam war. The war—a major boon for defense contractors—was winding down at the same time the Apollo program put a man on the moon. That, combined with an economic downturn greatly intensified by the OPEC oil embargo, hurt aerospace firms dramatically. Boeing Co., a Seattle, Washington-based aerospace firm, which could have been a major beneficiary of an SST research contract, was forced to shrink dramatically.

(A legendary joke of the time was that a billboard was posted along a highway leading away from Seattle, saying, "Would the last one out turn out the lights?"

The Question

One scientist peripherally involved with the SST debate, Sherry Rowland, came away from the meetings and discussions wondering just what was going on in the Earth's upper atmosphere. A presentation in 1970 by British scientist James Lovelock showed that there was a minute amount (230 parts per trillion, which has been compared to one drop in a large swimming pool) of CFCs in the lower atmosphere. Rowland, trained as a radiochemist (i.e., one who studies the effects of nuclear energy on chemicals), was no atmospheric scientist, but knew that cosmic radiation was a major element in the chemistry of the upper atmosphere, the shield that protects the Earth from the full force of solar energy. Atmospheric chemistry at the time was a little-understood subject, especially that involving the upper atmosphere. ("Weather" such as we obtain from a meteorological report is primarily the study of the lower atmosphere.) Only in the 1950s, when high-altitude balloons and small rockets could be sent into the upper atmosphere, were actual samples taken.

The existence of ozone in the upper atmosphere had been postulated some 40 years earlier, and was later found to be about three parts per thousand (0.3 percent), depending on altitude. Even this amount wasn't what it should be, based on purely chemical-reaction terms. Whatever the source of the ozone, scientists soon realized its beneficial effect: Because of its chemical structure, it could absorb and radiate back to space the ultraviolet radiation the sun emits. Not all of this radiation is blocked; it is the cause of suntans among beach bathers. It also causes skin cancer, and it interacts with plant and animal life in ways that are uncertain to this day.

Rowland, a middle-aged, tenured scientist at the University of California, decided that the effects of CFCs in the atmosphere would be a good research topic for one of his postdoctoral students. Mario Molina took the project on. After studying the literature and running calculations, Molina came to realize that, if the CFCs were getting to stratospheric levels, then they would dissociate under the solar radiation, producing "free" (that is, unbonded) chlorine. And with a start, in late 1973, Molina came to the conclusion that this free chlorine could react with ozone, reducing it to normal oxygen. The protective blanket surrounding the Earth would begin to tatter.

The procedure, among scientists, to make a momentous conjecture of this sort is to publish a paper, in as prestigious a journal as possible, and subject it to the scrutiny of the world scientific community. This occurred in mid-1974, in the British journal *Nature*. Among the most alarming conjectures that Molina and Rowland made in this paper was that a tremendous lag effect would take place between the time a CFC molecule is produced, when it reaches the upper atmosphere, and when, having destroyed more than 1,000 ozone molecules for each atom of free chlorine, it finally ceases to have an effect. For some of the more durable CFC forms, this lag would be a century. The CFCs that were theorized to be affecting the ozone layer were those, by and large, manufactured in the 1930s and 1940s. And if all CFC production were to cease today, the presence of CFCs in the atmosphere would not end until the twenty-second century.

The world did not sit up and take notice when the paper was published. Although a number of atmospheric scientists were very intrigued by the implications of the paper, the general scientific and business communities were being buffeted by a slew of reported, confirmed, or postulated environmental hazards: DDT in polar bears, crude-oil spills in oceans, and microbes from astronauts returning from the Moon. CFCs could be another one to chuck onto the pile. Enough interest was generated to convene a panel of reviewers from the National Academy of Sciences, a federally chartered, but independently managed, research body. At the same time, a number of grass-roots environmental groups, having latched onto the issue, recommended that at the very least, consumers should not use aerosol sprays containing CFCs. This application could well be a frivolous one, given the implications of the Rowland-Molina theory. In 1974, the Natural Research Defense Council, on the advice of its lawyers and scientists, recommended that the federal Consumer Product Safety Commission review the safety of CFCs in aerosols.

Bureaucracies being what they are, it wasn't until late in the year that a presidential task force was assembled to review the theory yet again and then, in mid-1975, recommended that the National Academy of Sciences conduct a full-scale study to resolve the issue. Congressional hearings were also held, but a bill to restrict the use of CFCs until the environmental risk was examined was not passed. Several states, allotted a full share of environmentally conscious legislators, considered and passed state bans on aerosol cans containing CFCs. By now, a set piece would occur whenever a state legislature considered CFC use. Rowland or another scientist would appear before a panel to testify, as would industry representatives of either the CFC industry or the aerosol can industry.

About this time, too, industry set up a Council on Atmospheric Sciences to develop its own research and technical expertise. Groups such as these usually allocate funds to conduct research within their own laboratories or those of universities. Invariably, they come up with results that serve the interests of their sponsors. One assumes that this is simply a case of squashing studies that have negative implications for the industry, while broadcasting those that have a positive result. Not necessarily "bad" science, but certainly limited science. At any rate, the research that such groups conduct is subjected to the same scientific rigor as any other study scientists conduct. If nothing else, these studies serve as an anvil on which other research can be pounded out.

By 1975, Russell Train, head of EPA, took the stand that international guidelines should be set up for CFC use, a proposal that went nowhere, given the general doubts that the rest of the world had concerning the Rowland-Molina theory. The American consumer, however, unilaterally decided to act. Sales of aerosol cans dropped substantially between 1973 and 1975. In mid-1975, Johnson Wax, a major consumer-goods manufacturer bolted from industry ranks and decided to stop manufacturing CFC-containing aerosol products. The announcement highlights the product testing and evaluation that goes on behind the scenes among most manufacturers. This manufacturer, whose engineers and marketing managers realized fairly quickly what a frivolous use of CFCs aerosol cans are, felt no pain in substituting another propellant.

First Reviews

The NAS panel, which had been working on the issue for months, was straining to the breaking point to announce something meaningful. As a variety of hypotheses were suggested and tested, the results indicated everything from a more rapid reduction to an actual increase in atmospheric ozone. Finally, in late 1976, the NAS panel's report came out, essentially validating the Rowland-Molina hypothesis. While the actual details were still unclear, the burden of evidence seemed to favor the stance that, indeed, CFCs could cause injury to the ozone layer. No atmospheric sampling or experimenting were done, but a great number of laboratory tests and computer simulations were run. As evidence of how disunited the panel was, however, it recommended that more study, and not any sort of restricted use, was necessary. Other government agencies, however, soon decided that the scientific evidence in the report was more compelling than the policy recommendations and moved to ban aerosol and other "nonessential" uses of CFCs. These agencies were the Food and Drug Administration and the Consumer Product Safety Commission; EPA joined in as well.

The aerosol can industry, having fought aggressively for two years to dispel worries over aerosol cans and to win back its shrinking market, quickly surrendered. Already, new product formulations or mechanical devices (such as a hand-pumped spray) were under development in industry laboratories. These appeared on the marketplace in short order. "Aerosol producers had had two years to anticipate a ban of the chemicals and spray cans and they responded with the kind of entrepreneurial spirit and technological skills that make America great," wrote Sharon Roan in *Ozone Crisis: The 15-year Evolution of a Sudden Global Emergency*.

The key qualifier to the ban, from industry's point of view, was the phrase "nonessential use." At the outset, HVAC and certain other applications were deemed "essential" and consumption could continue. There follows one of the typical clashes of perspective that many other pollution problems revolve around and that make sense only on a case-by-case basis. Within five or so years of the aerosol ban, total U.S. annual consumption of CFCs exceeded what had existed prior to the ban. Environmentalists look on this as a failure of regulatory policy—an exercising of a loophole, since the total amount of CFCs fated to reach the stratosphere would inevitably increase. Industrialists look on this as a lost market opportunity. Most products, under normal circumstances, increase in year-to-year consumption as population grows and economic activity increases. If there were no ban at all, CFC production would have been proportionately much, much greater. Other sections of this book mentioned similar results for such pollutants as coliform bacteria in surface water, lead vapor in air, or sulfur dioxides around city environments. All of these are produced in huge volumes, yet those volumes are dramatically smaller than earlier projections that were based on normal rates of growth. Who is the "winner" and who is the "loser" when deciding between these differing perspectives?

At any rate, there followed from 1980 to 1985 a slack period in the CFC debate. The Reagan administration, with aggressively anti-environmentalist policies, came into office committed to lifting or postponing as many environmental restrictions as it could. The anti-CFC "movement," if such a diverse group of scientists, nonprofit groups, and industrialists could be so characterized, lost momentum in the flush of the initial victory. Scientific studies to take samples of the upper atmosphere were carried out—some with industry funding. Laboratories, both in academia and in industry, worked on the chemistry of CFCs and ozone. Arrayed against this movement was the Alliance for Responsible CFC Policy, an agency funded primarily by the CFC producers but including some 500 members of the CFC-using industries: the HVAC manufacturers, the electronics industry, and the processors of polyurethane foams.

Around 1980, too, another, even more serious environmental controversy began to flare up with greater force: the "greenhouse effect." First proposed in the early 1960s, this theory is all but confirmed today, but its ultimate effects are still being researched. The greenhouse effect postulates that the increase of carbon dioxide and other gases in the atmosphere, mostly as a result of human activity, will cause a rise in the earth's temperature over the next century. (There are also many natural greenhouse gases of consequence in the matter.) CFCs, it is shown, acted as greenhouse gases as well as ozone depletors. A new set of scientists, including some of the same scientists studying CFCs, began to gather about this issue as the 1980s wore on.

Three more studies were released by the National Academy of Sciences during the 1979–1984 period, all indicating some ozone depletion, all indicating serious questions about the theory. Two new pieces of data, acquired under completely unexpected conditions, suddenly lent new drama to the debate in 1984. A European group, the British Antarctic Survey, found that ozone concentrations there plummeted by 40 percent during late fall in 1982, but the data were so unusual that they repeated the measurements over two more years to confirm them. When a paper was published (again, in *Nature*) on the result, other scientists at NASA's Goddard Space Flight Center realized that it confirmed data they had seen and rejected as spurious because they were so low. The Goddard data came from the Nimbus 7 satellite, which had been launched years earlier to conduct atmospheric research.

The identification of this "ozone hole" lent new credibility to the concerns over the entire ozone layer. Something—no one was sure what—caused the ozone to be eaten up inside the polar vortex that forms in the upper atmosphere at the end of the austral winter. (Picture water going down a bathtub drain: The space is still in the middle of the drain, even though water is swirling rapidly around it.) When the vortex breaks up as sunlight again reaches the southern pole, the hole dissipates.

On to the Antarctic

Now, a concerted effort got underway to study this hole. The research depended on the various research teams that go to the Antarctic every year, supplemented by a group of atmospheric scientists from the National Oceanic and Atmospheric Administration (NOAA). The Antarctic hole was seen to be a harbinger of future ozone depletion, and if the presence of CFCs and free chlorine could be found there, it would confirm what was believed to be occurring on a much smaller (and harder to measure) scale in the rest of the atmosphere. By 1986, EPA, under a new administrator, issued a proposed regulation to ban all CFCs, not just "nonessential" ones. The preliminary efforts to start an international negotiation to carry out this ban worldwide also began, in coordination with the United Nations Environmental Programme (UNEP). EPA had also, by now, organized a Stratospheric Ozone Program within its own staff.

Under National Science Foundation auspices, a flight over the Antarctic, timed for the appearance of the ozone hole, was planned for the fall of 1987. A key element was the development of suitable instruments that measure not just ozone but chlorine-containing compounds and other atmospheric gases. The instruments, normally in use in temperate rooms in U.S. laboratories, would also have to contend with the intense cold of the Antarctic. Both overflights, with the instruments on aircraft, and balloon launches were planned for this National Ozone Expedition (NOZE II, as it later came to be called; the previous year, NOZE I was run with ground-based instruments only). Weather was a further uncertainty. Many relatively untested techniques would be used; the scientists would be contending with the ability to get any results at all, let alone meaningful ones. Instrumentation designers from NOAA, NASA, Harvard, the State University of New York, and other organizations were involved. Pilots from Lockheed, the aerospace firm, would fly the aircraft. A division of ITT, ITT McMurdo, was the contractor for NASA with regular responsibility for maintaining the NASA station in the Antarctic, and the Chemical Manufacturers Association kicked in some funding.

By now, the overall picture was becoming clear. Something, probably CFCs, was destroying ozone in the Antarctic, and the hole didn't disappear until the polar vortex broke up. The total effect, over time, would be to reduce ozone throughout the stratosphere faster than natural processes could restore it. The EPA/UNEP negotiations moved forward, spurred by the realization that, if the ozone-CFC connection were confirmed, the problem would intensify for decades to come, regardless what was done at the moment. CFC manufacturers set their research laboratories back to work developing alternative products. Users of CFCs, depending on how far-sighted their managements were, began making contingency plans for alternative technologies. When an external event like a banned chemical or manufacturing method occurs (or, in a more typical example, the availability of them becomes limited because of a temporary shortage or drastic price increase), new technologies rapidly come out of the woodwork, spurred forward by entrepreneurially minded businesspeople who can capitalize on the new opportunity. On the other hand, there have been, and will be for as long as is legally possible, manufacturers and users that will make or use CFCs until the very last day they can. Only the regulatory clout of government agencies will stop them.

NOZE I and II confirmed the CFC-ozone connection. The fact that the hole occurred in the Antarctic was attributed to the formation of polar stratospheric clouds (PSCs) of ice. These PSCs serve as a catalyst for the reaction by enabling the form that chlorine usually takes when freed from the CFC, chlorine nitrate, to react on the surface of the icy particle to form a chlorine radical (that is, a charged, reactive atom). This radical can react directly with the ozone. Without the water, the chlorine remains relatively unreactive in the nitrate form. CFCs' primary harmful effect is to enable the chlorine to float up into the upper atmosphere where the ozone is; chlorine alone would be washed out of the atmosphere at lower altitudes. The existence of PSCs was not even known until the overflights were conducted. Mario Molina, now working in the Jet Propulsion Laboratory of NASA, reappeared with a crucial paper demonstrating this effect—a full 14 years after his initial studies had kicked off the entire crisis.

Events moved forward quickly by now. The Alliance for Responsible CFC Policy announced that it would support a production limit on CFCs. Within days, Du Pont Co., the world's largest producer (and an Alliance member), went one step farther and announced that it would begin limiting CFC production unilaterally and would offer alternatives as the marketplace dictated. Its management asked for "an orderly transition" developed by international protocol. For the previous few years, too, nonprofit organizations, such as Friends of the Earth, the Natural Resources Defense Council, and the Environmental Defense Fund aimed their publicity guns at CFC production, seeking restrictions. As the scientific data poured in, these campaigns had a much greater effect.

The Montreal Protocol

A draft resolution limiting CFC use and banning certain forms of CFCs altogether was written in the summer of 1987. An international conference in Montreal brought together 43 nations, including most of the industrialized nations that produce and consume the largest amounts of CFCs. The Protocol was signed in September. Two months later, the NOZE II results came in, validating nearly all points of the CFC-ozone connection. A follow-up UNEP conference for 1990 was scheduled at the time of the first one, and the NOZE II and other results led to an acceleration and further limitation of CFC production.

By the time of the second UNEP conference, the greenhouse-effect issue began to loom larger and larger. Many rather unnatural observations were spurring this debate forward, especially in the United States. California was entering the fourth straight year of drought. Two years earlier, in 1988, a nationwide heat wave, combined with a lack of rain, caused crop failures across the nation and dropped the Mississippi River to levels too low to support river traffic. As the 1980s drew to a close, new records for daily temperatures were being set all over the country, and when all the data were in, 1990 was found to be the warmest year of the century.

It is uncertain whether there is a cause-and-effect relationship between these heat waves and droughts and the greenhouse effect itself. James Hansen, a NASA scientist, asserts that the greenhouse effect is real and is happening now, but many other scientists feel that it is premature to draw that conclusion. Nevertheless, many European countries are considering "carbon dioxide taxes" on fuels and other sources of carbon dioxide to wean society away from its fossil-fuel dependence. While the United States had been leading the effort to restrict CFCs, it is currently retarding efforts to address the greenhouse issue.

In 1990, numerous new technologies were announced to compensate for the CFC ban. In the electronics industry, a joint effort between the Department of Defense, EPA, and an industry trade group led to the validation of an electronics-cleaning process employing a chemical called *terpene* (which is derived from pine trees and other plants) in place of CFCs. Terpene is not only as good as CFCs; it is also cheaper. Companies that manufacture the production machinery the electronics industry uses have already brought out new versions that can use the terpenes. In the polyurethane foam industry, chemists dusted off old processes that mixed water with the polyurethane chemicals prior to starting the solidification reaction. The water reacts to form carbon dioxide gas, which in turn provides the foaming effect. (Yes, carbon dioxide is a worry as a greenhouse gas, but the amount of it produced from polyurethane production is minuscule compared to burning oil, coal, and gas. Molecule for molecule, the CFCs are much more harmful than the carbon dioxide that replaces them.)

Industry Adjusts

In HVAC, the knottiest technical problems remain to be addressed by the mechanical engineers, physicists, and chemists who design these systems. The alternative CFCs that companies like Du Pont have put forth (with names like R-134a and R-141b) are not as energy-efficient as the banned

CFCs. In addition, they affect the lubrication system inside the refrigerator's compressor. New lubricants are in the offing.

HVAC manufacturers are being hit with a double whammy by the CFC ban. Starting in the 1970s, the U.S. Department of Energy began pushing for mandatory improvements in the energy efficiency of home appliances. One way to achieve this improvement is to use better insulation, and when manufacturers looked around for a suitable technology, the best they could find was to use CFC-blown polyurethane foam. In this case, the CFCs not only provide the right foam structure, but the chemical itself serves as an insulator when it is locked inside the foam structure (which, in turn, is locked inside the walls of the refrigerator cabinet). Banning CFCs in the polyurethane foam leads to more electricity demand, which leads to more coal burning, which leads to an increased greenhouse effect. One candidate technology, still in a very preliminary research stage, is to create a foam out of glass under vacuum conditions. This foam, extremely light and fragile, but an excellent insulator, is called an *aerogel*. It was developed by scientists at a Department of Energy laboratory.

There are those who argue, with some merit, that all CFC production should stop immediately, given the existing damage to the ozone layer. There are also those who would argue (as James Watt, Secretary of the Interior, did in the early 1980s) that the only necessary solutions were to wear hats on sunny days and use more sunscreen. The Montreal Protocol (as updated in 1990), using the evidence at the time, represented a negotiated settlement between these divergent views. As it stands, worldwide ozone depletion will be on the order of 10 percent toward the end of the twenty-first century. For every 1 percent decrease in ozone, by some estimates, there is a 2 percent increase in ultraviolet light and a comparable increase in skin cancer. CFCs are also felt to be approximately 25 percent of the causative agents of greenhouse warming, and thus will contribute a couple degrees of the expected 3- to 10-degree rise.

Du Pont, a central player in the entire CFC controversy, has stated that it will be spending more than $1 billion over the coming decade in researching alternative products and building factories to make them. The chemical industry as a whole will be spending upwards of $6 billion worldwide. The Alliance for Responsible CFC Policy estimates that 5,000 businesses, employing 700,000 workers, producing goods and services worth more than $28 billion per year, are connected to the use of CFCs. Industry will be making adjustments for years to come. And, since the United States represents about a third of worldwide CFC consumption, the worldwide value of CFC-related products is approximately $100 billion.

While a total, immediate ban is interesting to contemplate, the fact of the matter is that there are only 40-odd signatories to the Montreal Protocol; it leaves out much of the Third World. Literature references prior to the Tiennamen Square crisis in 1989 keep referring to a mysterious 1-to-3 million refrigerators that China was about to build for its people; how will they be manufactured? Also, after the Montreal Protocol was re-signed in 1990, a squabble developed over a multimillion-dollar fund to help the Third World adjust to a non-CFC future. Moreover, the total picture isn't in yet on the alternative refrigerants. An industry-sponsored program called PAFTT (Program for Alternative Fluorocarbon Toxicity Testing) is in the process of evaluating the health effects of the new compounds. The full range of tests takes five years and is complemented by a parallel set of tests for manufacturing and usage safety. In the meantime, the CFC industry, as well as a smattering of entrepreneurs, are targeting the CFC recycling business as a new opportunity. On the one hand, CFCs that are recycled are not released into the atmosphere; on the other, CFC molecules already in existence are as much of a threat to the ozone layer as the CFC molecule about to be manufactured in a chemical plant somewhere in the world.

Who the Players Were

The invention of CFCs was hailed in 1930 and damned in 1987. In the interim, a multibillion-dollar industry, employing hundreds of thousands of workers, came into being. How do green-collar workers deal with this industry and work force today? A look at the players involved in the CFC-ozone controversy gives some indication. The major groupings are:

■ The scientists, including atmospheric scientists, physicists, chemists, and medical specialists.

■ The engineers, including those building the instrumentation used to uncover and monitor the CFC problem, as well as those working to change current manufacturing technology and product designs in HVAC, electronics, and polyurethane plastics.

■ The business managers, who had to evaluate the rising CFC crisis, formulate responses, make contingency plans, and put those plans into effect.

■ The regulators and government officials who funded the research, evaluated policy options, and negotiated the ban and other trade agreements. The CFC-ozone controversy is notable among environmental

issues for being one of the first to result in an internationally negotiated treaty. More are expected to come in the future, calling to duty
those with training in international relations.

■ The nonprofit executives, ranging from scientists who performed independent analyses of industry and government data, to lawyers who
pursued legal actions to get government moving, to lobbyists who agi-
tated for new laws.

■ The communicators, working for government, industry, nonprofits,
and the scientific community. These included the editors and journalists who broadcast the information the scientists were developing to the
rest of the scientific and business community, the public affairs officers
of the nonprofits and of government agencies who deal with newspaper and television journalists, the public at large, and congressional
staffers.

Much debate has been generated over whether it should have taken a
total of 14 years, from 1973 to 1987, to confirm the threat and to take action
against it. One of the key knowledge gaps, it seems clear, was in atmospheric
science. One hesitates to contemplate how long the pro-and-con arguments
would have gone on without the Antarctic research station data from the
British or the confirming satellite data from the United States. Both these
sets of data were being gathered for pure-science motives—that is, science
for science's sake—rather than for direct results on the CFC debate.

On the industry side, it is conceivable, but highly doubtful, that scientists could have been employed to carry out the basic research in atmospheric chemistry. Industry does sponsor basic research, but more and more
of late this research is being tied to current or contemplated new technologies that could be commercialized. Once the potential threat of CFCs was
raised, it seems inexcusable that research into producing alternatives pro-
ceeded on such a desultory basis. Serious research to figure out how to produce alternative CFCs really got under way only in the mid-1980s, as the
ozone-hole discovery was being made. Industrialists argue that such research
is very expensive and is hard to justify when there is no imminent chance of
reward. But this was not a case of inventing a whole new technology. Many
of the alternative products were known to exist and had been synthesized in
small, laboratory-scale volumes for decades. The engineering and scientific
work to scale up the laboratory-sample production is an effort that goes on
continuously for a wide variety of products.

Among CFC users, it seems that the transition is going more quickly.
The switch to non-CFC propellants among aerosol-can producers was prac-

tically immediate. The transition is going surprisingly quickly in the case of electronics producers, who now seem to have a much improved cleaning process at hand in the use of terpenes. (Ironically, this technology, developed by a company called Petroferm, Inc., in Fernandina Beach, Florida, was already being developed to solve another environmental problem—the toxic sludges produced when oily metal parts are cleaned during manufacturing or repair work—according to Michael Hayes, the product manager at Petroferm.)

In many cases, the same companies that produce CFCs also produce polyurethane-plastic raw materials, so in this CFC market, a company is calling on the expertise of one of its divisions to help the needs of another division. One of the leading CFC firms, Imperial Chemical Industries of Great Britain (called ICI Americas in the United States, where it has extensive production facilities) pulled an unusual triple play in this regard. At around the same time that its CFC division was announcing commercial production of a new CFC alternative, its lubricants division was announcing a new lubricant that is compatible with its refrigerant, and its polyurethanes division unveiled new production methods for polyurethane foams that did not depend on CFCs. In general, it seems that polyurethane applications for everything except insulation are well on their way to CFC replacement. However, insulation is the single largest market for polyurethanes on a volume basis.

The HVAC industry has dozens of firms constantly modifying their products to gain incremental advantages in the marketplace, and the switch to alternative CFCs is mostly one other ingredient to throw into the technology-development stew. A number of entrepreneurial firms are even developing technologies that obviate the need for CFCs or alternative CFCs altogether by using water, air, helium, or other natural materials.

There is, however, a natural inertia to the HVAC industry, in that its products are expected to last for a decade and sometimes longer. Markets for such products as these demand a consistent level of service from their suppliers; one simply doesn't toss out the product every year and buy a new one. Thus, the purchasers of HVAC systems (including commercial building operators, other industries, and the homeowner, who also is likely to have an air-conditioned car) will need CFCs or a near-equivalent for years to come. Moreover, as the energy-efficiency-standard issue comes to the fore, it is evident that there is more than one criterion under which the environmental friendliness of a product can be judged. Finally, it is also clear that there are potential liabilities even in a product as ubiquitous as refrigerators: If a toxic material is introduced into them without a thorough evaluation of the risks inherent in that material, lives are at stake.

All industries, not just the CFC-using ones, have several powerful motivators for increasing the resources they devote to hiring engineers, scientists, and planners to evaluate the environmental risks and benefits of the businesses they are in. Besides the legal threat of a shutdown when something like a CFC ban is instituted, the product- or workplace-liability threat when lives are needlessly lost because of faulty manufacturing demands a new level of expertise. In certain instances, it is also now possible for a common citizen (who may also be a current or former employee of a firm) to sue that firm for unrevealed environmental damages. Finally, as a number of examples in earlier chapters showed, companies can gain a winning marketing edge by being "greener" than their competitors. As long as enough American consumers demand a higher level of environmental soundness in the products they buy, U.S. manufacturers will seek to improve their types of products and methods of manufacturing to meet those demands.

Of the major sectors of the green-collar work force that were described in Chapter 2, the teaching profession is the only one that is missing from the CFC-ozone story. I could argue, somewhat puckishly, that in fact it was a teacher (Sherry Rowland) who got the whole controversy started by guiding the research of his student, Mario Molina. There is, however, some more broad-based teaching going on behind the scenes: Du Pont, in announcing its new CFC alternatives (which will be marketed under the name Suva— bye, bye Freon!), also announced a greatly expanded CFC recycling campaign. In order to make HVAC system-maintenance personnel aware of the program and of its procedures, the company created a fairly elaborate interactive video and computer system for self-training. "Du Pont's CFC Conservation Training Program is an interactive video course for management, operators, and mechanics on how to operate and maintain industrial refrigeration equipment more effectively and efficiently to minimize the release of CFC refrigerants into the atmosphere," says company sales literature. That's just one example. The American Society of Heating, Refrigeration and Air-Conditioning Engineers, Inc. (ASHRAE) has a rather comprehensive set of lectures, books, and study materials that it sells to members during national conferences.

Perhaps the biggest teaching involvement with CFCs will be for school teachers at all levels to guide their charges through the thickets of environmental issues, using this CFC story as a cautionary tale. In this way, new generations of students will learn that, indeed, the sky was falling in the 1980s.

APPENDIX A

Nonprofit Organizations

Air and Waste Management
   Association
P.O. Box 2861
Pittsburgh PA 15230
(412) 232 3444

Alliance for Engineering in
   Medicine and Biology
1101 Connecticut Ave., NW
Washington, DC 20036
(202) 857 1199

Alliance for Environmental
   Education Inc.
10751 Ambassador Dr., Suite 201
Manassas, VA 22110
(703) 335 1025

American Academy of
   Environmental Engineers
132 Holiday Ct., Suite 206
Annapolis, MD 21401
(301) 266 3311

American Association for the
   Advancement of Science
1333 H St., NW
Washington, DC 20005
(202) 326 6400

American Chemical Society
1155 16th St., NW
Washington, DC 20036
(202) 872 4600

American Conference of
   Governmental Industrial Hygienists
6500 Glenway Ave., Bldg. D-7
Cincinnati, OH 45211
(513) 661 7881

American Council of Independent
   Laboratories
Suite 412
1725 K St., NW
Washington, DC 20006
(202) 887 5872

American Entomological Society
1900 Race St.
Philadelphia, PA 19103
(215) 561 3978

American Farmland Trust
1920 N St., NW, Suite 400
Washington, DC 20036
(202) 659 5170

American Fisheries Society
5410 Grosvenor Lane
Suite 110
Bethesda, MD 20814
(301) 897 8616

American Forestry Association
P.O. Box 2000
Washington, DC 20013
(202) 667 3300

American Industrial Hygiene
   Association
475 Wolf Ledges Parkway
Akron, OH 44311
(216) 762 7294

American Institute of Aeronautics
   and Astronautics
370 L'Enfant Promenade, SW
Washington, DC 20024
(202) 646 7400

American Institute of Architects
1735 New York Ave., NW
Washington, DC 20006
(202) 626 7300

American Institute of Chemical
   Engineers
345 East 47th St.
New York, NY 10017
(212) 705 7338

American Institute of Mining,
   Metallurgical and Petroleum
   Engineers (AIME)
345 East 47th St.
New York, NY 10017
(212) 705 7695

American Institute of Physics
335 East 45th St.
New York, NY 10017
(212) 661 9404

American Institute of Plant Engineers
3975 Erie Ave.
Cincinnati, OH 45208
(513) 561 6000

American Nuclear Society
555 North Kensington Ave.
La Grange Park, IL 60525
(708) 352 6611

American Planning Association
1776 Massachusetts Ave., NW
Washington, DC 20036
(202) 872 0611

American Public Health Association
1015 15th St., NW
Washington, DC 20005
(202) 789 5600

American Rivers
801 Pennsylvania Ave., SE
Washington, DC 20003
(202) 547 6900

American Society for Microbiology
1913 Eye St., NW
Washington, DC 20006
(202) 833 9680

American Society of Agricultural
   Engineers
2950 Niles Rd.
St. Joseph, MI 49085
(616) 429 0300

American Society of Agronomy
677 South Segoe Rd.
Madison, WI 53711
(608) 273 8080

American Society of Civil Engineers
Student Services Dept.
345 East 47th St.
New York, NY 10017
(212) 705 7496

American Society of Consulting
  Planners
210 7th St.
Washington, DC 20003

American Society of Heating,
  Refrigerating and
  Air-Conditioning Engineers, Inc.
  (ASHRAE)
1791 Tulie Circle, NE
Atlanta, GA 30329
(404) 636 8400

American Society of Mechanical
  Engineers
345 East 47th St.
New York, NY 10017
(212) 705 7722

American Society of Safety Engineers
1800 E. Oakton St.
Des Plaines, IL 60018
(708) 692 4121

American Water Resources
  Association
5410 Grosvenor Lane
Bethesda MD 20814
(301) 493 8600

American Water Works Association
6666 W. Quincy Ave.
Denver, CO 80235
(303) 794 7711

American Wilderness Alliance
7600 E. Arapahoe Rd., Suite 114
Englewood, CO 80112
(303) 771 0380

Association of American Geographers
1710 16th St., NW
Washington, DC 20009
(202) 234 1450

Association of Corporate
  Environmental Officers
P.O. Box 4117
Timonium, MD 21093
(800) 876 6618

Association of Environmental and
  Resource Economists
1616 P. St., NW
Washington, DC 20036
(202) 328 5000

Association of Ground Water
  Scientists and Engineers
6375 Riverside Dr.
Dublin, OH 43017
(614) 761 1711

Association of State and Interstate
  Water Pollution Control
  Administrators
444 N Capitol St., NW
Washington, DC 20001

Board of Certified Safety Professionals
208 Burwash Ave.
Savoy, IL 61874
(217) 359 9263

Center for Hazardous Materials
  Research
320 William Pitt Way
Pittsburgh, PA 15238
(412) 826 5320

Center for Marine Conservation
1725 DeSales St., NW
Washington, DC 20036
(202) 429 5609

Citizen's Clearinghouse for
  Hazardous Waste, Inc.
P.O. Box 926
Arlington, VA 22216
(703) 276 7070

Citizens for a Better Environment
942 Market St., No. 505
San Francisco, CA 94102
(415) 788 0690

Clean Water Action Project
317 Pennsylvania Ave., SE
Washington, DC 20003
(202) 547 1196

The Conservation Foundation
1250 24th St., NW
Washington, DC 20037
(202) 293 4800

Conservation International
1015 18th St., NW, Suite 1000
Washington, DC 20036
(202) 429 5660

Cousteau Society, Inc.
930 W. 21st. St.
Norfolk, VA 23517
(804) 627 1144

Defenders of Wildlife
1244 19th St., NW
Washington, DC 20036
(202) 659 9510

Earthwatch
P.O. Box 403N
Watertown, MA 02272
(617) 926 8200

Ecology Center
1403 Addison St.
Berkeley, CA 94702
(415) 548 2220

Entomological Society of America
9301 Annapolis Rd.
Lanham, MD 20706
(301) 731 4535

Environmental Action
1525 New Hampshire Ave., NW
Washington, DC 20036
(202) 745 4870

Environmental Defense Fund
257 Park Ave. South
New York, NY 10010
(212) 505 2100

Environmental Law Institute
1616 P St., NW, Suite 200
Washington, DC 20036
(202) 328 5150

Freshwater Foundation
2500 Shadywood Rd.
Navarre, MN 55392

Friends of the Earth
218 D St., SE
Washington, DC 20003
(202) 544 2600

Geological Society of America
3300 Penrose Place
Boulder, CO 80301
(303) 447 2020

Green Cross Certification Co.
1611 Telegraph Ave., Suite 1111
Oakland, CA 94612
(415) 832 1415

Greenpeace USA
1436 U St., NW
Washington, DC 20009
(202) 462 1177

Green Seal, Inc.
1733 Connecticut Ave., NW
Washington, DC 20009
(202) 328 8095

Hazardous Materials Control
  Research Institute
7237 Hanover Parkway
Greenbelt, MD 20770
(301) 982 9500

Institute for Local Self-Reliance
2425 18th St., NW
Washington, DC 20009
(202) 232 4108

Institute of Electrical and
  Electronics Engineers, Inc.
345 East 47th St.
New York, NY 10017
(212) 705 7900

Institute of Environmental Sciences
940 E. Northwest Highway
Mt. Prospect, IL 60056
(312) 255 1561

Institute of Industrial Engineers
25 Technology Park
Norcross, GA 30092
(404) 449 0460

Institute of Scrap Recycling Industries
1627 K St., NW
Washington, DC 20006
(202) 466 4050

International Association of
  Environmental Managers
243 W. Main St.
P.O. Box 308
Kutztown, PA 19530
(215) 683 5098

The Izaak Walton League of America
1401 Wilson Blvd., Level B
Arlington, VA 22209
(703) 528 1818

Junior Engineering Technical
  Society (JETS)
1420 King St.
Alexandria, VA 22314
(703) 548 5387

League of Conservation Voters
1150 Connecticut Ave., NW, Suite
  201
Washington, DC 20036
(202) 785 8683

National Association for
  Environmental Management
4400 Jenifer St., NW
Washington, DC 20015
(202) 966 0019

National Association of
  Environmental Professionals
P.O. Box 15210
Alexandria, VA 22309
(703) 660 2364

National Association of Professional
  Environmental Communicators
P.O. Box 06 8352
Chicago, IL 60606
(312) 781 1505

National Audubon Society
950 Third Ave.
New York, NY 10022
(212) 832 3200

National Environmental
  Development Association
1440 New York Ave., NW, Suite 300
Washington, DC 20005
(202) 638 1230

National Environmental Health
  Association
720 S. Colorado Blvd.
#970 South Tower
Denver, CO 80222
(303) 756 9090

National Recycling Coalition
1101 30th St., NW, Suite 305
Washington, DC 20007
(202) 625 6406

National Society of Professional
  Engineers
1420 King St.
Alexandria, VA 22314
(703) 684 2800

National Solid Wastes Management
  Association
1730 Rhode Island Ave., NW
Washington, DC 20036
(202) 659 4613

National Toxics Campaign
37 Temple Place, 4th Floor
Boston, MA 02111
(617) 482 1477

National Water Well Association
6375 Riverside Dr.
Dublin, OH 43017
(614) 761 1711

National Wildlife Federation
1400 16th St., NW
Washington, DC 20036
(202) 797 6800

Natural Resources Defense Council
40 West 20th St.
New York, NY 10011
(212) 727 2700

Nature Conservancy
1815 N. Lynn St.
Arlington, VA 22209
(703) 841 5300

New Alchemy Institute
237 Hatchville Rd.
East Falmouth, MA 02536
(508) 564 6301

North American Association for
  Environmental Education
P.O. Box 400
Troy, OH 45373
(513) 698 6493

Operations Research Society of
  America
Mt. Royal & Guilford Ave.
Baltimore, MD 21202
(301) 528 4146

Organic Crop Improvement
  Association
3185 Township Rd. 179
Bellafontaine, OH 43311
(513) 592 4983

Organic Food Alliance
2111 Wilson Blvd., Suite 531
Arlington, VA 22201
(703) 276 9498

Resource Policy Institute
P.O. Box 39185
Washington, DC 20016
(202) 895 2601

Sierra Club
730 Polk St.
San Francisco, CA 94109
(415) 776 2211

Society of Automotive Engineers
400 Commonwealth Dr.
Warrendale, PA 15096
(412) 776 4841

Society for Ecological Restoration
c/o University of Wisconsin
  Arboretum
1207 Seminole Highway
Madison, WI 53711
(608) 263 7889

Society of Environmental Toxicology
  and Chemistry
1101 14th St., NW, Suite 1100
Washington, DC 20005
(202) 371 1275

Society of Fire Protection Engineers
60 Batterymarch St.
Boston, MA 02110
(617) 482 0686

Society of Manufacturing Engineers
One SME Dr.
Dearborn, MI 48121
(313) 271 1500

Society of Plastics Engineers
14 Fairfield Dr.
Brookfield, CT 06804
(203) 775 0471

Society of Toxicology
1101 14th St., NW, Suite 1100
Washington, DC 20005
(202) 293 5935

Soil and Water Conservation Society
7515 N.E. Ankeny Rd.
Ankeny, IA 50021
(515) 289 2331

Soil Science Society of America
677 South Segoe Rd.
Madison, WI 52711
(608) 273 8080

State and Territorial Air Pollution
  Program Administrators and the
  Association of Local Air Pollution
  Control Officials
444 N. Capitol Street, NW
Washington, DC 20001
(202) 624 7864

Student Conservation Association,
  Inc.
P.O. Box 550
Charlestown, NH 03603
(603) 826 4301

Union of Concerned Scientists
26 Church St.
Cambridge, MA 02238
(617) 547 5552

Waste Watch
P.O. Box 39185
Washington, DC 20016
(202) 895 2601

Water Pollution Control Federation
601 Wythe St.
Alexandria, VA 22314
(703) 684 2400

The Wilderness Society
900 17th St., NW
Washington, DC 20006
(202) 833 2300

World Resources Institute
1709 New York Ave., NW
Washington, DC 20006
(202) 638 6300

Worldwatch Institute
1776 Massachusetts Ave., NW
Washington, DC 20036
(202) 452 1999

World Wildlife Fund
1250 24th St., NW
Washington, DC 20037
(202) 293 4800

Zero Population Growth
1400 16th St., NW
Suite 230
Washington, DC 20036
(202) 332 2200

APPENDIX B

Environmental Publications

Following is a list of publications about social issues, industrial practices, and some scientific research on environmental topics. I have excluded most academic journals in favor of those that are accessible to the general reader. At the same time, there is an emphasis on business publications. The main reason for this focus is to provide the publications that would be most useful to the job hunter; academic journals, while extremely valuable for environmental science, are written mostly for other scientists. Most of the business journals are not available on newsstands, and some of them actually restrict their subscriptions to those working in an industry. (You can usually get around this restriction by calling yourself a "consultant" on the subscription form.) Most large university libraries will have many of these publications on hand, as will many corporate libraries.

Amicus Journal
40 West 20th St.
New York, NY 10011
(212) 727 2700
(House organ of the Natural
   Resources Defense Council)

*BioCycle, the Journal of Waste
   Recycling*
The JG Press, Inc.
P.O. Box 351
18 South Seventh St.
Emmaus, PA 18049
(215) 967 4135

*Buzzworm, The Environmental
   Journal*
2305 Canyon Blvd., Suite 206
Boulder, CO 80302
(303) 442 1969

California Environmental News

Tri-State Environmental News

Texas Environmental News

Environmental News Network
760 Whalers Way, Suite 100-A
Fort Collins, CO 80525
(303) 229 0029

Chemical & Engineering News
1155 16th St., NW
Washington, DC 20036
(202) 872 4600
(House journal of the American
  Chemical Society)

Chemical Engineering
1221 Ave. of Americas, 43rd Fl.
New York, NY 10020
(212) 512 2000

Chemical Engineering Progress
American Institute of Chemical
  Engineers
345 East 47th St.
New York, NY 10017
(212) 705 7576
(House journal of the American
  Institute of Chemical Engineers)

Chemical Week
Chemical Week Associates
P.O. Box 1074
Southeastern, PA 19398
(215) 630 6380

Civil Engineering
American Society of Civil Engineers
345 East 47th St.
New York, NY 10017
(212) 705 7496

Clean Water Report
CIE Associates
237 Gretna Green Ct.
Alexandria, VA 22304

E, The Environmental Magazine
Earth Action Network, Inc.
28 Knight St.
Norwalk, CT 06851
(203) 854 5559

*Earth First! The Radical
  Environmental Journal*
P.O. Box 5871
Tucson, AZ 85703

Engineering News-Record
1221 Ave. of Americas
New York, NY 10020
(212) 512 2500

Environmental Business Journal
Environmental Business Publishing
  Inc.
827 Washington
San Diego, CA 92103
(619) 295 7685

Environmental Economics Journal
30 Springborn St.
Enfield, CT 06082

EPRI Journal
Electric Power Research Institute
P.O. Box 10412
Palo Alto, CA 94303
(415) 855 2000

*Environmental Science and
  Technology*
American Chemical Society
1155 16th St., NW
Washington, DC 20036
(202) 872 4600

Golub's Oil Pollution Bulletin
World Information Systems
P.O. Box 535 Havard Square Station
Cambridge, MA 02238

Graduating Engineer
Peterson's/COG Publishing Group
16030 Ventura Blvd., Suite 560
Encino, CA 91436
(818) 789 5293

Hazardous Materials Control
Hazardous Materials Control Institute
7237 Hanover Pky.
Greenbelt, MD 20770
(301) 982 9500

Hazmat World
Tower-Borner Publishing, Inc.
800 Roosevelt Rd.
Glen Ellyn, IL 60137

*In Business, the Magazine for
  Environmental Entrepreneuring*
The JG Press, Inc.
P.O. Box 323
18 South Seventh St.
Emmaus, PA 18049
(215) 967 4136

Inside EPA
301 G St., SW
Washington, DC 20024

The Journal of Conservation Biology
Society for Conservation Biology
3 Cambridge Center
Cambridge, MA 02141
(617) 225 0401

Journal of Environmental Health
National Environmental Health
  Association
720 S. Colorado Blvd.
Denver, CO 80222
(303) 756 9090

*Journal of Soil and Water
  Conservation*
Soil and Water Conservation Society
7515 NE Ankeny Rd.
Ankeny, IA 50021
(515) 289 2331

*Journal of the Water Pollution
  Control Federation*
601 Wythe St.
Alexandria, VA 22314

National Parks
National Parks and Conservation
  Association
1015 31st St., NW
Washington, DC 20007
(202) 944 8530

Natural History
P.O. Box 5000
Harlan, IA 51537
1 (800) 234 5252

New Age Journal
342 Western Ave.
Brighton, MA 02135
(617) 787 2005

Organic Times
New Hope Communication
1301 Spruce St.
Boulder, CO 80302

Pollution Engineering
1935 Shermar Rd.
Northbrook, IL 60062
(708) 498 9840

Pulp & Paper
500 Howard St.
San Francisco, CA 94105
(415) 397 1881

REAP Newsletter
Iowa Department of Natural
  Resources
Wallace State Office Bldg.
Des Moines, IA 50319

Resource Recovery Focus
National Solid Wastes Management
  Association
1730 Rhode Island Ave., NW
Washington, DC 20036
(202) 659 4613

Resource Recycling
Resource Recycling, Inc.
P.O. Box 10540
Portland, OR 97210
(503) 227 1319

Science
American Association for the
  Advancement of Science
1333 H St., NW
Washington, DC 20005
(202) 326 6400

Waste Tech News
131 Madison St.
Denver, CO 80206
(303) 394 2905

Water Resources Review
U.S. Geological Survey
MS 20
12201 Sunrise Valley Dr.
Reston, VA 22092
(703) 860 6127

Water Well Journal
Water Well Journal Publishing Co.
6375 Riverside Dr.
Dublin, OH 43017

Whole Earth Ecolog
available through bookstores or
Whole Earth Access
2990 Seventh Ave.
Berkeley, CA 94710
(800) 845 2000
(415) 845 3000

APPENDIX C

Federal Addresses

Most of the larger federal agencies have regional hiring offices in addition to the headquarters office in Washington. Also, the very largest federal agencies have hiring centers for divisions within the agency as well as the agency as a whole. Some of these divisional offices are listed here, under the title of the parent agency. The U.S. Environmental Protection Agency is listed first, including its ten regional offices. The publication *Career America*, from which all these addresses were drawn, lists most of the regional and divisional addresses and should be consulted for more details.

**ENVIRONMENTAL
PROTECTION AGENCY**

Headquarters Office

Recruitment Center
(PM-224)
401 M St., NW
Washington, DC 20460
1 (800) 338 1350
(202) 382 3305

Regional Offices

Regional Personnel Office
Environmental Protection Agency
Room 2203
John F. Kennedy Bldg.
Boston, MA 02203
(617) 565 3719
(CT, ME, MA, NH, RI, VT)

Regional Personnel Office
Environmental Protection Agency
Room 937-C
26 Federal Plaza
New York, NY 10278
(212) 264 0016
(NY, NY, PR, VI)

Regional Personnel Office
Environmental Protection Agency
841 Chestnut Bldg.
Philadelphia, PA 19107
(215) 597 9372
(DE, MD, DC, PA, VA, WV)

Regional Personnel Office
Environmental Protection Agency
345 Courtland St., NE
Atlanta, GA 30365
(404) 347 3486
(AL, FL, GA, KY, MS, TN, NC, SC)

Regional Personnel Office
Environmental Protection Agency
230 South Dearborn St.
Chicago, IL 60604
(312) 353 2026
(IL, IN, MN, MI, OH, WI)

Regional Personnel Office
Environmental Protection Agency
1445 Ross Ave.
Dallas, TX 75202
(214) 655 6560
(AR, LA, NM, TX, OK)

Regional Personnel Office
Environmental Protection Agency
726 Minnesota Ave.
Kansas City, KS 66101
(913) 236 2821
(IA, KS, MO, NE)

Regional Personnel Office
Environmental Protection Agency
One Denver Place
999 18th St., Suite 500
Denver, CO 80202
(303) 293 1487
(CO, MT, ND, SD, UT)

Regional Personnel Office
Environmental Protection Agency
215 Fremont St.
San Francisco, CA 94105
(415) 974 8016
(AZ, CA, NV, HW, Guam,
   American Samoa, Trust Territories,
   Wake Island)

Regional Personnel Office
Environmental Protection Agency
M/S 301
1200 Sixth Ave.
Seattle, WA 98101
(206) 442 2959
(ID, OR, WA, AK)

DEPARTMENT OF AGRICULTURE

Office of Personnel
Central Employment Unit
U.S. Department of Agriculture
Room 1080, South Building
Washington, DC 20250
(202) 447 5626

Agricultural Research Service

Personnel Division
Building 003, BARC-West
Beltsville, MD 20705
(301) 344 1124

Forest Service

Washington Office
P.O. Box 96090
Room 906, Rosslyn Plaza East
Washington, DC 20090-6090
(703) 235 2730

Soil Conservation Service

Personnel Division
P.O. Box 2890
Washington, DC 20013
(202) 447 2631

DEPARTMENT OF COMMERCE

This arm of government, with
   34,000 employees, has four regional
   "administrative support centers":

Personnel Officer
Eastern Administrative Support
  Center
253 Monticello Ave.
Norfolk, VA 23510
(804) 441 6516

Personnel Officer
Central Administrative Support
  Center
601 East 12th St.
Kansas City, MO 64106
(816) 758 2056

Personnel Officer
Mountain Administrative Support
  Center
325 Broadway
Boulder, CO 80303
(303) 497 6306

Personnel Officer
Western Administrative Support
  Center
7600 Sand Point Way, NE
BIN C15700
Seattle, WA 98115
(206) 526 6054

Bureau of the Census

Personnel Division
Room 3254, Building Three
Washington, DC 20233
(301) 763 5780

National Institute of Standards and Technology

Personnel Officer
Room A-123, Administration
  Building
Gaithersburg, MD 20899
(301) 975 3008

National Oceanic and Atmospheric Administration

Personnel Division
6010 Executive Blvd.
WSC #5, Room 706
Washington, DC 20852
(301) 443 8834

DEPARTMENT OF DEFENSE
DEPARTMENT OF THE ARMY

Army Corps of Engineers

Civilian Personnel Division
ATTN: CEPE-CS
20 Massachusetts Ave., NW
Room 5105
Washington, DC 20314-1000
(202) 272 0720

DEPARTMENT OF ENERGY

Headquarters Operations Division
Room 4E-090
1000 Independence Ave., SW
Washington, DC 20585
(202) 586 8536

DEPARTMENT OF HEALTH AND HUMAN SERVICES

Public Health Service

OASH Personnel Operations Office
5600 Fishers Lane
Room 17A-08
Rockville, MD 20857
(301) 443 6900

Agency for Toxic Substances
and Disease Registry

Personnel Office
1600 Clifton Rd., NE
Atlanta, GA 30333
(404) 639 3615

Centers for Disease Control

Personnel Office
1600 Clifton Rd., NE
Atlanta, GA 30333
(404) 639 3615

Food and Drug Administration

Division of Personnel Management
5600 Fishers Lane
Room 4B-41
Rockville, MD 20857
(301) 443 1970

National Institutes of Health
National Institute of
Environmental Health

Division of Personnel Management
9000 Rockville Pike
Bldg. 31, Room B3C15
Bethesda, MD 20205
(301) 496 2403

DEPARTMENT OF THE INTERIOR

Personnel Office
Office of the Secretary
Washington, DC 20240
(202) 343 6618

Bureau of Land Management

Division of Personnel
18th and C Sts., NW (MIB)
Washington, DC 20240
(202) 343 3193

Bureau of Mines

Headquarters
2401 E St., NW
Washington, DC 20241
(202) 634 4710

Bureau of Reclamation

Headquarters
18th and C Sts., NW
Washington, DC 20240
(202) 343 4626

National Parks Service

Headquarters
Branch of Personnel Operations
18th and C Sts., NW
P.O. Box 37127
Washington, DC 20013
(202) 343 4648

U.S. Fish and Wildlife Service

Headquarters
18th and C Sts., NW
Washington, DC 20240
(202) 343 6104

U.S. Geological Survey

National Center, MS-215
12201 Sunrise Valley Dr.
Reston, VA 22092
(703) 860 6127

DEPARTMENT OF LABOR

Occupational Safety and Health
Administration

Office of Personnel Management
Frances Perkins Bldg., Room N3308
200 Constitution Ave., NW
Washington, DC 20210
(202) 523 8013

NATIONAL AERONAUTICS AND SPACE ADMINISTRATION

Headquarters, DP
Washington, DC 20546
(202) 453 8480

NATIONAL SCIENCE FOUNDATION

Staffing Assistant
Division of Personnel and
  Management
1800 G St., NW, Room 208
Washington, DC 20550
(202) 357 9529

NUCLEAR REGULATORY COMMISSION

College Recruitment Coordinator
Office of Personnel
Washington, DC 20555
(301) 492 9027

APPENDIX D

State Agencies

ALABAMA

Conservation and Natural Resources
  Department
64 N. Union St.
Room 702
Montgomery, AL 36130
(205) 242 3486

ALASKA

Environmental Conservation
  Department
3220 Hospital Dr.
P.O. Box O
Juneau, AK 99811-1800
(907) 465 2606

ARIZONA

Environmental Quality Department
2005 N. Central Avenue
Phoenix, AZ 85004
(602) 257 2300

ARKANSAS

Health Department
4815 W. Markham
Little Rock AR 72201
(501) 661 2111

Pollution Control and Ecology
  Department
8001 National Dr.
Little Rock AR 72219
(501) 562 7444

CALIFORNIA

Environmental Affairs Agency
1102 Q St.
P.O. Box 2815
Sacramento, CA 95812

COLORADO

Natural Resources Department
1313 Sherman St.
Room 718
Denver, CO 80203
(303) 866 3311

CONNECTICUT

Environmental Protection
  Department
165 Capitol Ave.
Hartford, CT 06006
(203) 566 5599

DELAWARE

Natural Resources and
  Environmental Control
  Department
89 Kings Highway
P.O. Box 1401
Dover, DE 19903
(302) 736 4506

DISTRICT OF COLUMBIA

Public Works Department
2000 14th St., NW, 6th Floor
Washington, DC 20009
(202) 939 8000

FLORIDA

Environmental Regulation
  Department
2600 Blair Stone Rd.
Tallahassee, FL 32399-2400
(904) 488 9334

GEORGIA

Natural Resources Department
205 Butler Street, SE, Suite 1252
Atlanta, GA 30328
(404) 656 3530

HAWAII

Hawaiian Home Lands Department
P.O. Box 18789
Honolulu, HI 96805
(808) 548 6450

Environmental Health Administration
1250 Punchbowl St.
Honolulu, HI 96813
(808) 548 6455

IDAHO

Fish and Game Department
600 S. Walnut
P.O. Box 25
Boise, ID 83707
(208) 334 3782

ILLINOIS

Environmental Protection Agency
P.O. Box 19276
Springfield, IL 62794
(217) 782 3397

INDIANA

Environmental Management
  Department
105 S. Meridian St.
P.O. Box 6015
Indianapolis, IN 46206-6015
(317) 232 8162

IOWA

Natural Resources Department
Wallace Building
Des Moines, IA 50319-0034
(515) 281 5385

KANSAS

Health and Environmental
  Department
Forbes Field, Building 740
Topeka, KS 66620
(913) 296 1500

KENTUCKY

Natural Resources and
  Environmental Protection Cabinet
Capitol Plaza Tower, 5th Floor
Frankfort, KY 40601
(502) 564 2043

LOUISIANA

Environmental Quality Department
P.O. Box 44066
Baton Rouge, LA 70804
(504) 342 1222

MAINE

Environmental Protection
  Department State House
Station 17
Augusta, ME 04333
(207) 289 7688

MARYLAND

Natural Resources Department
Tawes State Office Building
Annapolis, MD 21401
(301) 974 3041

MASSACHUSETTS

Environmental Affairs Executive
  Office
100 Cambridge St., Room 2000
Boston, MA 02202
(617) 727 9800

MICHIGAN

Natural Resources Department
P.O. Box 30028
Lansing, MI 48909
(517) 373 1220

Pollution Control Agency
520 Lafayette Rd.
St. Paul, MN 55155
(612) 296 6300

MISSISSIPPI

Environmental Quality Department
P.O. Box 20305
Jackson, MS 39289-1305
(601) 961 5099

MISSOURI

Conservation Department
2901 W. Truman Blvd.
P.O. Box 180
Jefferson City, MO 65102-0180
(314) 751 4115

Natural Resources Department
P.O. Box 176
Jefferson City, MO 65102
(314) 751 3443

Public Safety Department
P.O. Box 749
Jefferson City, MO 65102
(314) 751 4905

MONTANA

Natural Resources and Conservation
  Department
1520 E. 6th Ave.
Helena, MT 59620-2301
(406) 444 6873

NEBRASKA

Environmental Control Department
State Office Building
P.O. Box 98922
Lincoln, NE 68509-8922
(402) 471 2186

NEVADA

Conservation and Natural Resources
   Department
201 S. Fall St.
Carson City, NV 89710
(702) 687 4360

NEW HAMPSHIRE

Environmental Services Department
6 Hazen Dr.
Concord, NH 03301
(603) 271 3503

NEW JERSEY

Environmental Protection
   Department
401 E. State St.
CN 402
Trenton, NJ 08625-0402
(609) 292 3131

NEW MEXICO

Health and Environment Department
1190 St. Francis Dr.
Santa Fe, NM 87503
(505) 827 0020

NEW YORK

Environmental Conservation
   Department
50 Wolf Rd.
Albany, NY 12233
(518) 457 3446

NORTH CAROLINA

Environment, Health and Natural
   Resources Department
P.O. Box 72687
Raleigh, NC 27611
(919) 733 4984

NORTH DAKOTA

Game and Fish Department
100 N. Bismarck Expressway
Bismarck, ND 58501
(701) 221 6300

Health and Consolidated
   Laboratories Department
600 E. Boulevard Ave.
Bismarck, ND 58505
(701) 224 2370

Parks and Recreation Department
1424 W. Century Ave., Suite 202
Bismarck, ND 58501
(701) 224 4887

OHIO

Environmental Protection Agency
1800 Watermark
P.O. Box 1049
Columbus, OH 43266-0149
(614) 644 3020

OKLAHOMA

Health Department
1000 NE 10th St.
P.O. Box 53551
Oklahoma City, OK 73152
(405) 271 4200

Public Safety Department
3600 Martin Luther King Ave.
P.O. Box 11415
Oklahoma City, OK 73136-0415

Wildlife Conservation Department
P.O. Box 53465
Oklahoma City, OK 73152

OREGON

Fish and Wildlife Department
P.O. Box 59
Portland, OR 97207
(503) 976 6339

PENNSYLVANIA

Environmental Resources
  Department
P.O. Box 2063
Harrisburg, PA 17120
(717) 783 2300

RHODE ISLAND

Environmental Management
  Department
9 Hayes St.
Providence, RI 02908
(401) 277 6800

SOUTH CAROLINA

Health Environmental Control
  Department
2600 Bull St.
Columbia, SC 29201
(803) 734 4880

SOUTH DAKOTA

Water and Natural Resources
  Department
Joe Foss Building
523 E. Capitol
Pierre, SD 57501
(605) 773 3151

TENNESSEE

Health and Environment Department
344 Cordell Hull Building
Nashville, TN 37219-5402
(615) 741 3111

TEXAS

Air Control Board
6330 Highway 2901 E.
Austin, TX 78723
(512) 451 5711

Health Department
1100 W. 49th St.
Austin, TX 78756
(512) 458 7244

Parks and Wildlife Department
4200 Smith School Rd.
Austin, TX 78744
(512) 389 4800

Soil and Water Conservation Board
P.O. Box 658
Temple, TX 76503
(817) 773 2250

Water Commission
P.O. Box 13087
Capitol Station
Austin, TX 78711
(512) 463 5538

Water Development Board
P.O. Box 13231
Austin, TX 78711-3231
(512) 463 7847

UTAH

Health Department
P.O. Box 16700
Salt Lake City, UT 84116-0700
(801) 538 6101

VERMONT

Natural Resources Agency
103 S. Main St.
Waterbury, VT 05676
(802) 244 6916

VIRGINIA

Natural Resources Secretariat
9th St. Office Building, Room 525
Richmond, VA 23219
(804) 786 0044

WASHINGTON

Ecology Department
MS PV-11
Olympia, WA 98504
(206) 459 6000

WEST VIRGINIA

Air Pollution Control Commission
1558 Washington St., E.
Charleston, WV 25311-2599
(304) 348 4022

Natural Resources Department
1900 Kanawha Boulevard E.
Charleston, WV 25305
(304) 348 2754

WISCONSIN

Natural Resources Department
P.O. Box 7921
Madison, WI 53707
(608) 266 2621

WYOMING

Environmental Quality Department
Herschler Building
122 W. 25th St., 4th Floor
Cheyenne, WY 82002
(307) 777 7937

PUERTO RICO

Environmental Quality Board
P.O. Box 11488
Santurce, PR 00910
(809) 725 5140

APPENDIX E

Educational Organizations

This section has two parts: the first contains a list of colleges and universities with environmental studies programs, emphasizing environmental engineering, at the graduate level. There are some 250 schools that have environmental studies at the undergraduate level; consult a directory such as *Peterson's Guide* or others if you are interested in them. The second part, noncredit ("certificate") programs, covers a range of nondegreed programs offered by schools, professional organizations, and private companies for training in the environmental field. Many of these programs are required for workers, managers, or trainers who perform environmental remediation projects such as at Superfund sites. Typically, an employer will send its staff people to these seminars (some of which run only a couple of days) in order to qualify for bidding on, and performing, government-funded or -licensed cleanup work. There is nothing that prevents the individual, on his or her own, from taking these courses in anticipation of a job, and they could prove beneficial as an indicator of deep interest in performing environmental work.

SCHOOL/INSTITUTION	PROGRAM
Credit Programs	
California Institute of Technology Division of Engineering and Applied Science Pasadena, CA 91125 (818) 356 6811	M.S., Ph.D. option in environmental engineering science

California Polytechnic State
University
Department of Civil and
Environmental Engineering
San Luis Obispo, CA 93407
(805) 756 0111

M.S. in environmental
engineering

Clemson University
Department of Environmental
Systems Engineering
Clemson, SC 29634
(803) 656 3311

M.Eng., M.S., Ph.D. in
environmental engineering

Colorado School of Mines
Department of Environmental
Sciences and Engineering
Ecology
Golden, CO 80401
(303) 273 3000

M.S. in ecological engineering

Colorado State University
Department of Civil Engineering
Fort Collins, CO 80523
(303) 491 1101

M.S., Ph.D. in environmental
engineering and water quality

Columbia University
School of Engineering and Applied
Science
Department of Chemical
Engineering and Applied
Chemistry
New York, NY 10027
(212) 854 1754

M.S., Eng.Sci.D., Ph.D. in
environmental control
engineering

Cornell University
Graduate Field of Engineering
Field of Civil and Environmental
Engineering
Ithaca, NY 14853
(607) 255 2000

M.Eng., M.S., Ph.D. in civil and
environmental engineering or
water resource systems

Pennsylvania State University
College of Engineering
Department of Civil Engineering
212 Sackett Bldg., Box E
University Park, PA 16802
(814) 865 4700

M.Eng., M.S., Ph.D. in
environmental engineering

Rensselaer Polytechnic Institute
School of Engineering and Science
Department of Environmental
  Engineering and Environmental
  Science
Troy, NY 12180
(518) 276 6000

M.Eng., M.S., Ph.D. in
environmental engineering

Rice University
Department of Environmental
  Science and Engineering
Houston, TX 77251
(713) 527 8101

M.E.E., M.E.S., M.S., Ph.D. in
environmental science or
engineering

Rutgers, The State
  University of New Jersey
Program in Civil and
  Environmental Engineering
New Brunswick, NJ 08093
(201) 932 1766

M.S., Ph.D. in environmental
engineering

State University of New York/
  Syracuse
College of Environmental Science
  and Forestry
Syracuse, NY 13210
(315) 470 6599

M.S., Ph.D. in environmental and
resource engineering

University of Alaska/Anchorage
School of Engineering
Program in Environmental
  Quality Engineering and
  Environmental Quality Science
Anchorage, AK 99508
(907) 786 1800

M.S., Ph.D. in science and in
engineering

University of California at Berkeley
School of Engineering and
  Applied Science
Department of Civil Engineering
Los Angeles, CA 94720
(415) 642 6000

M.Eng., M.S., Ph.D. in earthquake
engineering, geotechnical
engineering, water resources,
and environmental engineering

University of California at
  Los Angeles
School of Public Health
Program in Environmental
  Science and Engineering
Los Angeles, CA 90024
(213) 825 4321

D.Env.

University of Cincinnati
College of Medicine and Division
  of Graduate Studies and Research
Department of Environmental
  Health
(513) 556 6000

M.S., Ph.D. in environmental
hygiene and engineering

University of Colorado at Boulder
College of Engineering and
  Applied Science
Boulder, CO 80309
(303) 492 0111

M.S., Ph.D. in environmental
engineering, geotechnical
engineering, and water resources

University of Florida
College of Engineering
Department of Environmental
  Engineering Sciences
Gainesville, FL 32611
(904) 392 3261

M.E., M.S., Ph.D. in environmental engineering science

University of Houston
Program in Environmental
  Engineering
Houston, TX 77204
(713) 749 2321

M.S. Env. E., Ph.D. in
environmental engineering

University of Illinois at
  Urbana-Champaign
Department of Civil and
  Environmental Engineering
Urbana, IL 61801
(217) 333 1000

M.S., Ph.D. in environmental
engineering and science

University of Lowell
Program in Environmental Studies
1 University Ave.
Lowell, MA 01854
(508) 452 5000

M.Eng. in environmental studies

University of North Carolina at
  Chapel Hill
School of Public Health
Department of Environmental
  Sciences and Engineering
(919) 962 2211

M.S.E.E., M.S., Ph.D. in
environmental science

University of Oklahoma
School of Civil Engineering and
  Environmental Science
Norman, OK 73019

M.S., Ph.D. in water resources,
groundwater management,
environment science, occupational
safety and health, and others

University of West Virginia
College of Graduate Studies
Program in Environmental Studies
Institute, WV 25112
(304) 766 2044

M.S., Ph.D. in environmental
engineering

University of Wisconsin–Madison
Department of Civil and
  Environmental Engineering
Madison, WI 53706

M.S., Ph.D. in environmental
engineering

Washington State University
College of Engineering and
  Architecture
Department of Civil and
  Environmental Engineering
Pullman, WA 99164
(509) 335 3564

M.S., Ph.D. in environmental,
geotechnical, hydraulic, and other
engineering specialties

Wayne State University
Department of Chemical and
  Metallurgical Engineering
Detroit, MI 48202
(313) 577 3802
	M.S. in hazardous waste
	  management

Noncredit "Certificate" Programs

American Conference of
  Governmental Industrial
  Hygienists
6500 Glenway Ave.
Bldg. D-7
Cincinnati, OH 45211
(513) 661 7881
	Industrial hygiene

American Institute of Hazardous
  Materials Management
900 Isom Rd., Suite 103
San Antonio, TX 78216
(800) 729 6742
(512) 340 7775
	RCRA, OSHA compliance,
	  emergency response, and others

Center for Environmental
  Management
Tufts University
Curtis Hall
474 Boston Ave.
Medford, MA 02155
(617) 381 3531
	Asbestos and lead abatement,
	  environmental management, and
	  others

Center for Hazardous Materials
  Research
University of Pittsburgh
Applied Research Center
320 William Pitt Way
Pittsburgh, PA 15238
1 (800) 334 2467
(412) 826 5320
	Waste site health and safety,
	  emergency response, industrial
	  spills, hazardous waste
	  management, and others

Chemical Safety Associates
9163 Chesapeake Drive
San Diego, CA 92123
(619) 565 0302

Emergency response, laboratory
procedures

Con-Test Educational Center
Training/Marketing Coordinator
39 Spruce St.
East Longmeadow, MA 01208
(413) 525 1198

OSHA training

Delaware Technical and
   Community College
89 Christiana Rd.
New Castle, DE 19720
(302) 323 9602

Training-the-trainer
   environmental courses

Dennison Environmental Training
   Center
74 Commerce Way
Woburn, MA 01801
(617) 932 9400

Asbestos abatement, risk
   management, hazards
   communications, and others

Energy Support Services, Inc.
P.O. Box 6098
Asheville, NC 28816
(704) 258 8888

Hazardous materials
   management

The Environmental and
   Occupational Safety and
   Health Education and
   Training Center (EOSHETC)
Brookwood Plaza II
45 Knightsbridge Rd.
Piscataway, NJ 08854
(201) 463 5062

Comprehensive continuing
   education in environmental
   safety and health

The Environmental Institute
350 Franklin Rd.
Suite 300
Marietta, GA 30067
(404) 425 2000

Asbestos, lead abatement,
   environmental assessments,
   and related topics

Georgia Tech Research Institute
Georgia Institute of Technology
Atlanta, GA 30332
(800) 325 5007
(404) 894 7430

Environmental audits,
environmental safety and
health, and others

Geraghty & Miller, Inc.
c/o American Ecology Services, Inc.
127 East 59th St.
New York, NY 10022
(212) 371 1620

OSHA training for hazardous
wastes

Harvard School of Public Health
Office of Continuing Education
677 Huntington Ave.
Boston, MA 02115
(617) 432 3515

Risk analysis, environmental
safety and health, and others

Hazardous Materials Control
  Institute
Graduate Program
9300 Columbia Blvd.
Silver Spring, MD 20910
(301) 587 9390

Emergency response, hazardous
materials and waste
management, hydrogeology,
and others (runs both a
certificate and a degree
program in conjunction with
Wayne State University)

HAZMAT TISI (Training,
  Information and Services, Inc.)
Columbia Business Center
6480 Dobbin Rd.
Columbia, MD 21045

Hazardous materials training

Institute for Environmental
  Education
208 W. Cummings Park
Boston, MA 01801
(617) 935 7370

Asbestos and lead abatement,
training-the-trainer, and
others

National Environmental Training
  Association
8687 E. Via de Ventura
Suite 214
Scottsdale, AZ 85258
(602) 951 1440

Hazardous materials and waste
management, safety and health,
water treatment, emergency
response, and others

NUS Training Corporation
P.O. Box 6032
910 Clopper Rd.
Gaithersburg, MD 20877
(301) 258 2500

Hazardous communications,
safety, emergency response,
hazardous materials handling,
and others, on video formats

Rocky Mountain Center for
    Occupational and Environmental
    Health
Department of Family &
    Preventive Medicine
Bldg. 512
University of Utah
Salt Lake City, UT 84112
(801) 581 5710

Occupational safety and health,
industrial ergonomics, asbestos
sampling, industrial hygiene, and
others

Texas A&M University System
Texas Engineering Extension
    Service
College Station, TX 77843
(409) 845 6681

OSHA and EPA courses

University of California-Berkeley
    Extension
Programs in Environmental
    Hazard Management (PEHM)
2223 Fulton St.
Berkeley, CA 94720
(415) 643 7143
(Most other locations in the
    University of California
    system offer similar courses.)

Environmental inspection and
training and others

University of Cincinnati Medical
    Center
Institute of Environmental
    Health
Kettering Laboratory
3223 Eden Ave.
Cincinnati, OH 45267
(513) 558 1731

Asbestos abatement,
training-the-trainer, and others

University of Findlay
The Emergency Response Training
  Center
Hazardous Materials Management
  Program
1000 N. Main St.
Findlay, OH 45840
(419) 424 4647

OSHA training, environmental
  management, hazardous
  materials, and others

University of Kansas
Division of Continuing Education
6600 College Blvd., Suite 315
Overland Park, KS 66211
(913) 491 0221

Hazardous waste site training and
  others

University of Texas at Arlington
Center for Environmental
  Research and Training (CERT)
P.O. Box 19021
Arlington, TX 76019
(817) 273 3694

Asbestos abatement, hazardous
  wastes, emergency response,
  and others

University of Wisconsin-
  Madison/Extension
Department of Engineering
Professional Development
432 N. Lake St.
Madison, WI 53706
(800) 462 0876

Industrial pollution control,
  undergound storage tank
  management, hazardous waste
  site remediation, and others

Virginia Commonwealth University
Department of Preventive
  Medicine
Box 212 MCV Station
Richmond, VA 23298
(804) 786 9785

Environmental auditing,
  inspection, underground
  storage tanks, and others

References

American Society of Agronomy. *Exploring Careers in Agronomy.* Madison, WI: Author, 1989.

American Society of Heating, Refrigerating, and Air-Conditioning Engineers, Inc. *CFCs: Time of Transition.* New York: Author, 1989.

Baldwin J. (Ed.). *Whole Earth Ecolog.* New York: Harmony Books, 1990.

Berenyi, E., and Gould, R. *The 1991 Resource Recovery Yearbook.* New York: Government Advisory Associates, 1991.

Carson, Rachel. *Silent Spring.* Boston, MA: Houghton Mifflin, 1987.

Congressional Office of Technology Assessment. *Orbiting Debris: A Space Environmental Problem.* Washington, D.C.: Author, 1990.

DeAngelis, L. (Ed.). *The Complete Guide to Environmental Careers.* Washington, D.C.: Island Press, 1989.

Foreman, Dave. *Confessions of an Eco-Warrior.* New York: Harmony Books, 1991.

Fortuna, R. C., and Lennett, D. J. *Hazardous Waste Regulation: The New Era.* New York: McGraw-Hill, 1987.

Hammond, A. (Ed.). *World Resources 1990-91.* New York: Oxford University Press, 1990.

Hardin, Garrett. *Filters Against Folly.* New York: Viking Penguin, 1985.

McKibben, Bill. *The End of Nature.* New York: Random House, 1989.

Melosi, Martin. *Garbage in the Cities: Refuse, Reform, and the Environment, 1880-1980.* College Station, TX: Texas A&M University Press, 1981.

Miller, Richard K. *Environmental Markets, 1990-91.* Madison, GA: Richard K. Miller & Associates, Inc., 1991.

Openchowski, Charles. *A Guide to Environmental Law in Washington, D.C.* Washington, D.C.: Environmental Law Institute, 1990.

Roan, Sharon. *Ozone Crisis: The 15-Year Evolution of a Sudden Global Emergency.* New York: John Wiley & Sons, 1989.

U.S. Bureau of the Census. *Statistical Abstract of the United States, 1990.* Washington, D.C.: U.S. Government Printing Office, 1990.

Index

A

Agriculture Research Service, 76
Agronomist, 103
Air and Waste Management
    Association, 11, 23
Air quality engineer, 104
Alliance for Responsible CFC
    Policy, 149
American Chemical Society, 140
American Institute of Architects
    (AIA), 86
American Planning Association, 118
American Society of Civil
    Engineers, 23
American Society of Heating,
    Ventilating and
    Air-Conditioning, Inc., 154
Anderson, Dan, 90
Army Corps of Engineers, 11, 39
Atlantic Ridgefield Co., 62
Atomic Energy Commission, 52
Aveda Corporation, 62

B

Baruch, S.B., 38
Battelle Memorial Institute, 85
Bechtel National, Inc., 93

Berkibile, Robert, 53, 86
Best Available Technology (BAT), 43
Best Conventional Technology
    (BCT), 43
Best Practicable Technology (BPT),
    42
Bhopal, 49
Bierman-Lytle, Paul, 63
"Big Green," 94
Bio-Recovery Systems, 85
Biohazards specialist, 114
Biologist, 104
Biostatistician, 105
Boeing Co., 141
Brice, Gail, 6, 90
British Antarctic Survey, 146
Bureau of Census, 31
Bureau of Economic Analysis
    (BEA), 31, 32
Bureau of Land Managment
    (BLM), 37, 79
Bureau of Mines, 78
Bureau of Reclamation, 39, 78
Bush, George, 18, 25, 37

C

California Air Resources Board, 80

Carbon dioxide taxes, 149
Carothers, Andre, 62
Carson, Rachel, 12
CEIP Fund, 129
Chemical engineer, 106
Chemical Manufacturers
    Association, 22, 98
Chemical Waste Management, Inc.,
    46
Chemist, 106
Chesapeake Bay Foundation, 48
Civil engineer, 107
Clean Air Act (CAA), 4, 14, 25,
    37, 40, 41
Clear Water Act (CWA), 42, 43
Club of Rome, 15
Community relations manager, 107
Comprehensive Emergency
    Response, Compensation, and
    Liability Act (CERCLA), 17, 47
Congressional Office of Technology
    Assessment (OTA), 19
Conservation Foundation, 23
Consumer Product Safety
    Commission, 143
Contract administrator, 109

D

Database manager, 108
Dayton Engineering Laboratory, 139
Delaney, Maureen, 75
Demos, Edward, 3, 96
Dirksen, Everett, 33
Donohue & Associates, 93, 126
Du Pont Co., 55, 140, 149–150, 154

E

Earth First!, 24, 136
Earth scientist, 110
Ecologist, 110
Ecolo-Haul, 67

Endangered Species Act, 39
Energy Research and Development
    Commission, 52
Environmental Defense Fund, 13,
    22, 59, 98
Environmental engineer, 107
Environmental Impact Statement
    (EIS), 39, 40
Environmental manager, 102
Environmental Outfitters, 62
Environmental protection specialist,
    111
Environmental Resources
    Management Group, Inc., 82
"ETEX 91," 132

F

Federal Land Policy and
    Management, 37
Federal Trade Commission, 64
Federal Water Pollution Control
    Act, 42
Foreman, Dave, 24, 136
Forester, 111
Fortuna, R.C., 44
Frost & Sullivan, Inc., 33, 85, 86
Fund raiser, 112

G

General Schedule (GS), 74
Geological engineer, 113
Glitsch, Inc., 86
Government Advisory Associates, 69
Grass-roots coordinator, 113
Green-collar work, 2, 4, 19
Green Cross Certification Co., 63
Green Seal Inc., 63

H

Hardin, Garrett, 6
Hart Environmental Management
    Corp., 32

Hazardous materials specialist, 114
Hazardous Waste Treatment
    Council, 22
Heating, ventilating, and
    air-conditioning (HVAC), 139,
    145, 149–150, 152–153
Hooker Chemical Co., 16
Howson, Craig, 93

I

Industrial hygienist, 115
International Association for the
    Prevention of Smoke, 11
Interpreter, 116

J

Johnson, Mark, 90
Johnson Wax, 144
Junior Street-Cleaning League, 11

K

Kellogg, 11
Krishnaiah, Ravi, 82

L

Lash, Jonathan, 73, 88
Lawyer, environmental, 116
Lennett, D. J., 44
Lobbyist, 117
Lorenz, William, 45
Love Canal, 16, 31, 47, 48, 57
Lovelock, James, 85, 142
Lublin, J., 61

M

Management Information Services,
    Inc.(MIS), 34, 35, 36
McDonald's, 64
McKibben, Bill, 127
Melosi, Martin, 11
Midgley, Thomas, 139

Miller, Richard K., 66
Mitchell, Kathy, 88
Molina, Mario, 142, 154
Montreal Protocol, 148, 151
Muir, John, 10
Mynen, Ron van, 60

N

National Academy of Sciences, 143
National Ambient Air Quality
    Standards (NAAQS), 40
National Association for
    Environmental Management, 24
National Association of
    Environmental Professionals, 24
National Audubon Society, 12
National Environmental Policy Act
    (NEPA), 13, 39, 40
National Institute of Standards and
    Technology (NIST), 76
National Institutes of Health, 77
National Oceanic and Atmospheric
    Administration (NOAA), 77, 146
National Ozone Expedition
    (NOZE), 147, 148
National Parks Service, 78
National Pollutant Discharge
    Elimination System (NPDES),
    42
National Priority List (NPL), 48, 72
National Science Foundation (NSF),
    79
National Wildlife Federation, 12, 129
Natural Resources Defense Council,
    13, 143
Nature Conservancy, 95
New Source Performance Standards
    (NSPS), 40
Noise-control specialist, 118
North American Association for
    Environmental Education, 97
Not In My Back Yard (NIMBY),
    45, 46, 69

Nuclear Regulatory Commission
    (NRC), 79

O

Occupational Safety and Health Act
    (OSHA), 13, 14
Occupational Safety and Health
    Administration, 79
Openchowski, Charles, 48
Organization of Petroleum
    Exporting Countries (OPEC),
    14, 15
Ozone, 142
Ozone "hole," 146

P

Petersen, Gary, 67
Planner, 118
Pollution Abatement and Control
    (PAC), 31, 34, 35
Polyurethane foam, 141
Potentially Responsible Party (PRP),
    48, 49, 50
Powers, Ann, 48
Premanufacturing Notification
    (PMN), 47
Procter & Gamble, 62
Public Health Service, 77
Publicly Owned Treatment Works
    (POTW), 42, 43

R

Rathje, William, 68
Reilly, William, 23
Reith, Stephanie, 93, 126
Resource Conservation and
    Recovery Act (RCRA), 16, 43,
    44, 45, 46, 48, 49, 52
Richard K. Miller & Associates, 32,
    66
Rifkin, Jeremy, 40

Risk manager, 119
Rivers and Harbors Act, 11, 42
Rowland, Sherry, 142, 154

S

Safe Drinking Water Act (SDWA), 43
Safety engineer, 120
Sanitary Movement, 10
Schlatter, John, 92
Sherry & Bellows, 88
Sick building syndrome, 87
Sierra Club, 10, 22, 24
Soil scientist, 120
Space debris, 19
Spiszman, Jackie, 91
State Employment Service, 130
State Implementation Plans (SIP), 40
Superfund Amendments and
    Reauthorization Act (SARA),
    48, 49
Superfund Innovative Technology
    Evaluation (SITE), 50
Supersonic transport, 141
"Suva" refrigerants, 154
SVP, 32
Szuch, Clyde, 89, 117

T

Technical writer, 121
Tennessee Valley Authority, 11
Terpene, 149
Toxic Substance Control Act, The
    (TSCA), 47
Toxicologist, 122
Train, Russel, 144
Trans-Alaska Pipeline, 39
Trust for Public Land, 95
Turner, Jackson Frederick, 10

U

Union Carbide Corporation, 17, 49,
    60

United Nations Environmental
  Programme, 147, 149
U.S. Army Corps of Engineers, 77
U.S. Department of Agriculture, 76
U.S. Department of Commerce, 76
U.S. Department of Energy (DOE),
  15, 16, 31, 52
U.S. Department of Health and
  Human Services, 77
U.S. Department of the Interior, 37,
  78
U.S. Environmental Protection
  Agency (EPA), 3, 4, 13, 14,
  16, 23, 31, 32, 44, 45, 47, 48,
  49, 64, 65
    founding, 72
    structure, 75
U.S. Fish and Wildlife Service, 78
U.S. Food and Drug Administration,
  47
U.S. Forest Service, 76
U.S. Geological Survey, 78
U.S. Office of Personnel
  Administration (OPM), 130
U.S. Office of Technology
  Assesment, 31
U.S. Soil Conservation Service, 76

Urban Ore, Inc., 67

V

Venture capital firms, 132
Volunteerism, 128

W

Waring, George, 10
Waste Management, Inc., 67
Water Pollution Control Act, 14
Water Pollution Control Federation,
  23
Water quality technologist, 123
Watt, James, 24
Wellman, Inc., 65
Wilderness Society, 12
Wilderness Act, 37
Woods, Robin, 6
World Wildlife Fund, 23

Y

Yeager, K.E., 38
Yellowstone, 10
Yosemite Valley, 10